CONTENTS

MISSILE DEFENSE: A NEW ORGANIZATION, EVOLUTIONARY TECHNOLOGIES AND UNRESTRICTED TESTING

HEARING

BEFORE THE

SUBCOMMITTEE ON NATIONAL SECURITY, VETERANS AFFAIRS AND INTERNATIONAL RELATIONS

OF THE

COMMITTEE ON GOVERNMENT REFORM

HOUSE OF REPRESENTATIVES

ONE HUNDRED SEVENTH CONGRESS

SECOND SESSION

JULY 16, 2002

Serial No. 107–110

Printed for the use of the Committee on Government Reform

Available via the World Wide Web: http://www.gpo.gov/congress/house
http://www.house.gov/reform

U.S. GOVERNMENT PRINTING OFFICE

81–892 PDF WASHINGTON : 2003

For sale by the Superintendent of Documents, U.S. Government Printing Office
Internet: bookstore.gpo.gov Phone: toll free (866) 512–1800; DC area (202) 512–1800
Fax: (202) 512–2250 Mail: Stop SSOP, Washington, DC 20402–0001

COMMITTEE ON GOVERNMENT REFORM

DAN BURTON, Indiana, *Chairman*

BENJAMIN A. GILMAN, New York
CONSTANCE A. MORELLA, Maryland
CHRISTOPHER SHAYS, Connecticut
ILEANA ROS-LEHTINEN, Florida
JOHN M. McHUGH, New York
STEPHEN HORN, California
JOHN L. MICA, Florida
THOMAS M. DAVIS, Virginia
MARK E. SOUDER, Indiana
STEVEN C. LATOURETTE, Ohio
BOB BARR, Georgia
DAN MILLER, Florida
DOUG OSE, California
RON LEWIS, Kentucky
JO ANN DAVIS, Virginia
TODD RUSSELL PLATTS, Pennsylvania
DAVE WELDON, Florida
CHRIS CANNON, Utah
ADAM H. PUTNAM, Florida
C.L. "BUTCH" OTTER, Idaho
EDWARD L. SCHROCK, Virginia
JOHN J. DUNCAN, Jr., Tennessee
JOHN SULLIVAN, Oklahoma

HENRY A. WAXMAN, California
TOM LANTOS, California
MAJOR R. OWENS, New York
EDOLPHUS TOWNS, New York
PAUL E. KANJORSKI, Pennsylvania
PATSY T. MINK, Hawaii
CAROLYN B. MALONEY, New York
ELEANOR HOLMES NORTON, Washington, DC
ELIJAH E. CUMMINGS, Maryland
DENNIS J. KUCINICH, Ohio
ROD R. BLAGOJEVICH, Illinois
DANNY K. DAVIS, Illinois
JOHN F. TIERNEY, Massachusetts
JIM TURNER, Texas
THOMAS H. ALLEN, Maine
JANICE D. SCHAKOWSKY, Illinois
WM. LACY CLAY, Missouri
DIANE E. WATSON, California
STEPHEN F. LYNCH, Massachusetts

———

BERNARD SANDERS, Vermont (Independent)

KEVIN BINGER, *Staff Director*
DANIEL R. MOLL, *Deputy Staff Director*
JAMES C. WILSON, *Chief Counsel*
ROBERT A. BRIGGS, *Chief Clerk*
PHIL SCHILIRO, *Minority Staff Director*

SUBCOMMITTEE ON NATIONAL SECURITY, VETERANS AFFAIRS AND INTERNATIONAL RELATIONS

CHRISTOPHER SHAYS, Connecticut, *Chairman*

ADAM H. PUTNAM, Florida
BENJAMIN A. GILMAN, New York
ILEANA ROS-LEHTINEN, Florida
JOHN M. McHUGH, New York
STEVEN C. LATOURETTE, Ohio
RON LEWIS, Kentucky
TODD RUSSELL PLATTS, Pennsylvania
DAVE WELDON, Florida
C.L. "BUTCH" OTTER, Idaho
EDWARD L. SCHROCK, Virginia

DENNIS J. KUCINICH, Ohio
BERNARD SANDERS, Vermont
THOMAS H. ALLEN, Maine
TOM LANTOS, California
JOHN F. TIERNEY, Massachusetts
JANICE D. SCHAKOWSKY, Illinois
WM. LACY CLAY, Missouri
DIANE E. WATSON, California
STEPHEN F. LYNCH, Massachusetts

EX OFFICIO

DAN BURTON, Indiana
HENRY A. WAXMAN, California

LAWRENCE J. HALLORAN, *Staff Director and Counsel*
R. NICHOLAS PALARINO, *Senior Policy Advisor*
JASON CHUNG, *Clerk*
DAVID RAPALLO, *Minority Counsel*

MISSILE DEFENSE: A NEW ORGANIZATION, EVOLUTIONARY TECHNOLOGIES AND UNRESTRICTED TESTING

TUESDAY, JULY 16, 2002

HOUSE OF REPRESENTATIVES,
SUBCOMMITTEE ON NATIONAL SECURITY, VETERANS
AFFAIRS AND INTERNATIONAL RELATIONS,
COMMITTEE ON GOVERNMENT REFORM,
Washington, DC.

The subcommittee met, pursuant to notice, at 10 a.m., in room 2154, Rayburn House Office Building, Hon. Christopher Shays (chairman of the subcommittee) presiding.

Present: Representatives Shay, Putnam, Gilman, Platts, Schrock, Kucinich, Schakowsky, Tierney, Allen, and Lynch.

Staff present: Lawrence J. Halloran, staff director and counsel; J. Vincent Chase, chief investigator; Dr. R. Nicholas Palarino, senior policy advisor; Jason M. Chung, clerk; David Rapallo, minority counsel; and Earley Green, minority assistant clerk.

Mr. SHAYS. Good morning, everyone. I'd like to welcome our witnesses, to both panels, our guests, and obviously I welcome all our Members.

Under the National Missile Defense Act of 1999, it is the policy of the United States to deploy a missile defense system as soon as technically possible. Today we continue our oversight of the complex process of assessing and managing technological possibilities.

Almost 2 years ago, the subcommittee heard from the Department of Defense [DOD], the Department of State and others expressing a wide diversity of views on the technical feasibility of the national missile defense envisioned by the Clinton administration. Some believe technical progress had been hobbled by treaty constraints. Others foresaw a rush to failure as the political drive to deploy a missile shield ignored scientific uncertainties.

Since then, the program has been refocused to include boost phase systems like the airborne laser [ABL], missile defense office has been elevated to a DOD agency, the U.S. withdrawal from the anti-ballistic missile [ABM], treaty took effect last month. In view of these changes, Representative Tierney, a very active member of the subcommittee, asked that we revisit the process of judging the readiness of missile defense technologies.

That process has raised issues about the articulation of system requirements, the testing agenda, program structure, cost controls and deployment thresholds. Not at issue, the reality and immediacy of the threat.

Last April, subcommittee members received a classified briefing on national missile programs threatening regional and global security. That briefing confirmed a key finding of the Commission to Assess the Ballistic Missile Threat to the United States, also known as the Rumsfeld Commission, that, "The threat to the U.S. posed by these emerging capabilities is broader, more mature and evolving more rapidly than had been reported in estimates and reports by the intelligence community."

In securing our Nation in a volatile world, ranking threats and vulnerabilities is as essential as avoiding illusionary choices. Terrorists, acting as human delivery systems, do not need missiles to bring chemical, biological or even nuclear weapons to this continent. But that fact alone should not blind us to the emerging peril posed by nations developing and proliferating missile technologies. It is no coincidence that the roster of terrorism state sponsors contains many of the same nations building offensive missile capabilities.

Before the threats emerge, each much be confronted on its own terms. So it is not a question of whether, but when the United States begins to deploy a baseline missile defense system against known hostile capabilities. That timing will be driven by the legal mandate to do so as soon as possible, and by a knowledge base development and acquisition process that will add new technologies to the layered missile defense system envisioned by the administration.

It is that process we examine this morning. Our witnesses bring impressive breadth of experience and depth of expertise to our discussion of these important issues, and we look forward to their testimony.

I might add that I was someone who was an opponent of this missile defense, and someone who believes that it has become a necessity. Having said that, I'm someone who is determined that we not move forward until we actually have a system that is workable. And I would also say to you that I think we have a panel that allows both those who are critical, skeptical, even opponents of the system to be able to gather some very important information. So I think we're going to learn a lot from this hearing, and I thank all of our witnesses.

[The prepared statement of Hon. Christopher Shays follows:]

ONE HUNDRED SEVENTH CONGRESS

Congress of the United States

House of Representatives

COMMITTEE ON GOVERNMENT REFORM

2157 RAYBURN HOUSE OFFICE BUILDING

WASHINGTON, DC 20515–6143

MAJORITY (202) 225–5074
FACSIMILE (202) 225–3974
MINORITY (202) 225–5051
TTY (202) 225–6852

www.house.gov/reform

SUBCOMMITTEE ON NATIONAL SECURITY, VETERANS AFFAIRS,
AND INTERNATIONAL RELATIONS
Christopher Shays, Connecticut
Chairman
Room B-372 Rayburn Building
Washington, D.C. 20515
Tel: 202 225-2548
Fax: 202 225-2382
GROC.NS@mail.house.gov
http://www.house.gov/reform/ns/

Statement of Rep. Christopher Shays
July 16, 2002

Under the National Missile Defense Act of 1999, it is the policy of the United States to deploy a missile defense system as soon as "technologically possible." Today we continue our oversight of the complex process of assessing and managing technological possibilities.

Almost two years ago, the Subcommittee heard from the Department of Defense (DOD), the Department of State and others expressing a wide diversity of views on the technical feasibility of the National Missile Defense envisioned by the Clinton Administration. Some believed technical progress had been hobbled by treaty constraints. Others foresaw a "rush to failure" as the political drive to deploy a missile shield ignored scientific uncertainties.

Since then, the program has been refocused to include boost phase systems like the Airborne Laser (ABL). The missile defense office has been elevated to a DOD agency. United States withdrawal from the Anti-Ballistic Missile (ABM) Treaty took effect last month.

Statement of Rep. Christopher Shays
July 16, 2002
Page 2 of 2

In view of these changes, Rep. Tierney (MA), a very active Member of the Subcommittee, asked that we revisit the process of judging the readiness of missile defense technologies. That process has raised issues about the articulation of system requirements, the testing agenda, program structure, cost controls and deployment thresholds.

Not at issue: the reality and immediacy of the threat. Last April, Subcommittee Members received a classified briefing on national missile programs threatening regional and global security. That briefing confirmed a key finding of the Commission To Assess the Ballistic Missile Threat to the United States (also known as the "Rumsfeld Commission") that, "The threat to the U.S. posed by these emerging capabilities is broader, more mature and evolving more rapidly than has been reported in estimates and reports by the Intelligence Community."

In securing our nation in a volatile world, ranking threats and vulnerabilities is as essential as avoiding illusory choices. Terrorists, acting as human delivery systems, do not need missiles to bring chemical, biological or even nuclear weapons to this continent. But that fact alone should not blind us to the emerging peril posed by nations developing and proliferating missile technologies. It is no coincidence the roster of terrorism's state sponsors contains many of the same nations building offensive missile capabilities. Before the threats merge, each must be confronted on its own terms.

So it is not a question of whether, but when the United States begins to deploy a baseline missile defense system against known hostile capabilities. That timing will be driven by the legal mandate to do so as soon as possible, and by a knowledge-based development and acquisition process that will add new technologies to the layered missile defense system envisioned by the Administration. It is that process we examine this morning.

Our witnesses bring impressive breadth of experience and depth of expertise to our discussion of these important issues. We look forward to their testimony.

Mr. SHAYS. At this time I would recognize our distinguished Ranking Member, Mr. Kucinich.

Mr. KUCINICH. I want to thank the Chairman for calling this hearing regarding missile defense to explore a solution which is unworkable and unaffordable for a problem which is undefinable and often not believable. I'm very concerned about the recent actions by the Defense Department to reorganize its missile defense programs. Although the Pentagon has argued that the reorganization was necessary to speed up development of missile defenses, I'm concerned it will significantly reduce the oversight of and attention to the program's enormous costs and technical challenges.

On January 2, 2002, Secretary of Defense Rumsfeld issued a directive reorganizing the missile defense programs at the Pentagon. The change redesignated the Ballistic Missile Defense Office as the Missile Defense Agency. The new emphasis defined all of the missile defense initiatives as one large research and development program. This action reduces the oversight required by Congress.

Operational requirement documents were eliminated relating to individual programs. Time lines for development will not be established, and the Department of Defense has declined to set an overall architecture for this new system.

I might say that under these circumstances, the possibility for the taxpayers to be cheated is pretty serious. The DOD calls this new evolutionary acquisition strategy, in which the development, testing and acquisition all could occur simultaneously, rather than waiting for an entire system to be proven to work, the Pentagon could examine discrete successes in research and development at regular intervals and decide whether some aspect of the technology might serve a useful military purposes. Pentagon officials could take what they like and put it in the field. They could upgrade it later if the technology developed further.

In this spiral development strategy, the Pentagon will not focus on strict requirements for the program, but on various capabilities that may develop. From an oversight perspective, this approach has very few parameters by which to judge success. I think that's intentional. There is no way Congress can determine whether the program is over budget, because there are no cost projections. There is no way to determine an appropriate level of funding, because unlike other major weapons development programs, there is no final goal to which the program is striving.

There is also no way for Congress to gauge the appropriate pace of the program. Congress cannot determine whether progress is slow, because there are no dates by which the requirements must be met. In fact, there are no requirements at all. So what does this say about Congress's Constitutional oversight role?

In effect, what DOD is trying to do is eliminate Congress's oversight role while it spends, according to CBO estimates, at least $238 billion on a program that everyone at this point with any honesty knows doesn't work. Although the Pentagon has provided cost estimates for each element prior to the reorganization, it has since refused to provide a cost estimate for the ballistic missile defense system. This is because the Pentagon claims it does not know what the system will be.

Indeed. But without cost projections for the system, it's impossible for Congress to determine whether the program is over budget.

This concern is significant, because Pentagon weapons programs routinely exceed their initial projections, often by substantial amounts. That's how contractors get the business. Indeed, in addition, it's impossible for Congress to make risk benefit decision that place the missile defense system in the context of all other weapons systems. It's also difficult to know exactly what the American taxpayers would get for their dollar. Defense? Unlikely.

Although the actions of the Pentagon in restructuring the missile defense program are purportedly designed to speed up the process of technology development and deployment of militarily useful assets, I maintain another goal is to reduce oversight and prevent cost delays and technological flaws from garnering public attention. There is evidence that the Pentagon will go to somewhat extended lengths to prevent public disclosure of negative aspects of its program.

One example is a report by Mr. Coyle that listed numerous flaws in the testing program. This unclassified report was prepared by the director of the independent Pentagon testing office created by Congress specifically to provide an unbiased review of testing adequacy. In a hearing in September 2000 before this subcommittee, Mr. Coyle testified regarding the contents of this report, and pursuant to a request by Representative Tierney, agreed to provide it. The Department of Defense tried repeatedly to keep the Coyle Report from Congress. Despite numerous requests from Mr. Tierney, the Pentagon refused to deliver the report for over 8 months, disregarding a statute requiring the report be provided.

The Pentagon finally delivered the report after 55 Members of Congress, including the ranking minority members of three congressional committees, wrote to the Secretary of Defense. Even then, the Pentagon's official position remained, the public should be denied access to this unclassified information. In addition to failing to produce the report, the Missile Defense Agency had begun to withhold other information that was previously available.

Last month, Congress was informed of an abrupt oral directive from General Kadish. He ordered his deputies to classify as secret all information relating to decoys, even if general in nature, despite the fact that this information has been available for decades. In an op ed on June 11, 2002, Mr. Coyle, who has since left Government service, wrote "Some 20 developmental tests, each costing $100 million, will be needed before the ground based mid-course defense program is ready for the next step realistic operational testing. It may be the end of this decade before such testing with real world decoys can begin."

Thus the current test program is not giving away any secrets, nor is there any danger of that for years to come. The new classification policy is not justified by either the progress of tests so far or by the realism of the test. So what we have here, Mr. Chairman, is an effort by the Department of Defense to eliminate congressional oversight. And I might say, it's congressional oversight on an issue that's very serious, the defense of our country.

And we're leaving the American people to believe that somehow they're going to be protected against missiles which may in some distant future come in from some undefined enemy that has undefined technology. So far, all I see with this missile defense program is a bunch of baloney.

Mr. SHAYS. I thank the gentleman.

At this time, the Chair recognizes the vice chairman of the committee, Mr. Putnam.

Mr. PUTNAM. Thank you, Mr. Chairman. Thank you for putting this hearing together and your leadership on this issue.

I had not prepared an opening statement, instead looking forward to hearing from the expert testimony.

I do find it somewhat interesting that those who have expressed their passionate opposition to the program and to the spiral development process are in essence holding up the current procurement and acquisition process as a model of efficiency. This is, the reference to an undefined enemy and an undefined technology I think is somewhat inaccurate, in that we do have a defined enemy, a list of nations which are rapidly developing the technology and in some cases, have developed the technology to successfully launch short and medium range missiles and are well on their way toward developing long range missiles which threaten our troops in theater and within the near future, the continental United States.

So I would respectfully disagree with my colleague in that we do have a defined enemy, we do have defined technologies. We have had briefings and hearings on those nations and on those technologies. And this subcommittee has in its oversight role had access to that information. So I think that to a degree, you sell the oversight role of the Congress and your own abilities and oversight and this subcommittee short in that we have exercised that right, we have had reports forthcoming and had GAO prepare additional reports. And we are here today to take testimony from those people who are involved in that.

So from the oversight capacity, I would respectfully disagree that there is a malicious conspiracy at the Pentagon to withhold information from this subcommittee. With that, I thank the chairman and look forward to the testimony.

Mr. SHAYS. I thank the gentleman.

At this time, the Chair recognizes the distinguished gentleman from Massachusetts, Mr. Tierney.

Mr. TIERNEY. I thank the chairman.

I want to express my gratitude to Chairman Shays for conducting today's hearing and want to extend my appreciation to the panelists for the time, the insight and their testimony.

I do want to make a comment to my colleague from Florida, I don't think anybody is holding up what's gone on in the past as a model of efficiency. But I think it has been a model of transparency in oversight. One of the reasons we're having this hearing today is, if this procedure continues to go forward as the way that the Pentagon has proposed it, I fear and others fear that you can kiss that goodbye. You won't be having meaningful oversight hearings any more because you won't have any method by which to measure progress or costs as things go on.

That's why I have serious concerns about the way the missile defense program has been reorganized under this administration. Whether we agree with the program or don't agree with the program, I think our responsibility as a Congress is to make sure that if one is going forward, that we have an idea of what the costs are, that we have an idea of what its efficiencies are and its capabilities are. And it concerns billions of dollars. I don't think any of us want that wasted. And we don't want it thrown away as we proceed down this course.

When the Secretary issued his pronouncement on January 2, that he was reorganizing the programs, he made fundamental changes that I think threaten our ability to have oversight. They are probably going to threaten our ability to have any informed choices about the Nation's priorities, and they're going to reduce the involvement of the independent offices of Mr. Christie and others, taking away from them the opportunity to evaluate testing, establish military requirements and predict costs.

The directive virtually combined all the existing missile defense programs, regardless of the stages of development, into a single new ballistic missile defense system. Incredibly enough, a number of those systems had been at a stage where they were being evaluated, and this method of clumping them all together sets them all back to a point where we're now supposed to believe that there's just no opportunity to set out specific goals and achievement markers.

It's also removed formal military requirements from the programs. It then was converted into a giant research and development program with no parameters to gauge success. So there's no architecture, there's no time line, there's no cost estimate. Essentially there are no limits, and we all ought to be concerned about that. We ought to be concerned when Secretary Rumsfeld makes the statement, and I suspect he was serious, but I'm disturbed about the fact that he was, his statement in July was, we don't have a system, we don't have an architecture, we don't have a proposed architecture. All we have is a couple of handfuls of very interesting research and development and testing programs.

Well, you know, for $238 billion of projected costs, you ought to have a hell of a lot more than that. I think it's insulting to Congress and insulting to the American people to project forward that this is where they're at and we're all supposed to take it on good faith. This unprecedented level of flexibility may be appropriate, I don't know that it is, but it may be appropriate for some minuscule program, but surely not when you're dropping $8 billion a year in research for the most technologically daunting weapons program ever attempted.

When we met last in September 2000, the subcommittee heard horror stories about the ground based midcourse system. With respect to cost, we discovered that they were skyrocketing. In fact, General, I remember you estimating about $21 billion, and it wasn't until we showed you one of your own internal memos that you had to acknowledge it was already up to $63 billion.

With respect to that whole situation, their independent cost estimating group that were provided, now you have no yardstick to gauge success. So even though you estimated, you estimated wrong, at least we had some measurement on which to go forward. Now we have no way of knowing whether the program measures up or not.

With respect to capabilities, we heard even more dire descriptions. We heard from the Pentagon Chief Independent Testing Evaluator, Mr. Coyle, Mr. Christie's predecessor, who told us that testing in the program was severely deficient. It didn't test against basic threats, and it was so immature it could not be evaluated in terms of potential deployment dates.

Again, we had yardsticks, we had an operational requirements document and a testing evaluation master plan. Now, after reorganization, we have none of those tools.

On costs, the Pentagon refuses to provide any estimates for its new system, which is very convenient, even though Defense officials are pushing ahead with the same ground based midcourse system this subcommittee referred 2 years ago, they refuse to provide a cost estimate. In this case, how are we to know whether the system goes over budget? How are we going to know if the funds are being wasted? How will we make tough choices on whether to spend funds on missile defense or other priorities, such as bioterrorism preparedness? And the answer is, we cannot.

Just yesterday the administration came out with what was supposed to be its threat and risk assessment. In fact, they don't even mention anything other than homeland security issues, and don't stack up the different threats that this country faces one against the other at all.

We have the same problem with requirements in testing in this national missile defense program. The directive eliminated operational requirements documents, converting the program into a vast research and development project. Despite a year and a half in office, the administration has yet to develop a test and evaluation master plan to describe a specific test, with specific goals and the time lines that it will conduct.

That document was originally due in June. Now we're told it won't be completed until the fall. So how do we evaluate whether the Pentagon is meeting its goals? How do we know the program is progressing efficiently? How do we know if the program is even worth pursuing?

The rationale for this incredible flexibility is that urgency demands it. I think Mr. Miller is going to talk a little bit later about how many times in history we have heard urgency demands this kind of an approach, only to find out that we spend more money and get farther behind.

Rather than designing a rational developmental program that specifies concrete testing goals and provides comprehensive budget estimates, the Pentagon wants to throw everything they have into building something as soon as possible. That's why the administration pulled out of the ABM treaty prematurely, and that's why the administration is lurching headlong toward building missile interceptor silos in Alaska that have not been proven, cannot be fired in tests, and will provide absolutely no protection by 2004, notwithstanding the administration's numerous promises.

On this score, I ask unanimous consent to submit for the record, Mr. Chairman, the transcript of a special investigative briefing I held on June 11th discussing how the administration has made these promises of protection in 2004 by ignoring the technology and rushing ahead into deployment.

Mr. SHAYS. Without objection, so ordered.

Mr. TIERNEY. Thank you.

As I mentioned a minute ago, Mr. Miller is going to discuss in the second panel that history is littered with examples of weapons programs that were designed too quickly, that were deployed without sufficient testing and have suffered from fatal flaws, the Sergeant York gun, the Bradley fighting vehicle, the B–1 bomber and the C–5 cargo plane. These programs were rushed because of threats that were thought to be urgent at the time. They all unnecessarily cost American taxpayers billions of dollars to retroactively fix problems that should have been addressed earlier.

Mr. Chairman, I will close by noting that the Pentagon claims reorganization is not intended to reduce oversight. Even though there are no cost estimates, no operational requirements documents, no testing plan and no ultimate architecture, the Pentagon argues that congressional oversight will be just as vigorous as it has been in the past. With all due respect, this subcommittee in particular has to be wary of those types of claims.

As you know, the Pentagon ignored this subcommittee for over 8 months last year when it refused to deliver an unclassified report from Mr. Coyle describing in detail the flaws in the ground based test program. Even though a Federal statute required the Pentagon to deliver that report, the Pentagon wanted to hide those embarrassing results. It wasn't until 55 Members of Congress, including the ranking member and members of three congressional committees, wrote to Secretary Rumsfeld that we finally received it. Even then the Pentagon's official position remained that the public should be denied access to this report, even though its contents were completely unclassified.

As a result, the Missile Defense Agency's new strategy is to simply classify more information. Last month, we learned of a directive from General Kadish ordering the classification of all information relating to decoys, even if general in nature.

Mr. Coyle, who now writes from beyond Government service, confirmed that this action was uncalled for. This is what he had to say, "It may be the end of this decade before such testing with real world decoys can begin. Thus the current test program is not giving away any secrets, nor is there any danger of that for years to come. The new classification policy is not justified by either the progress in tests so far or by the realism of the tests. If this secrecy is not justified by the progress or the realism of the tests, it can only be explained by the Pentagon's desire to hide problems with progress and realism."

This kind of behavior, Mr. Chairman, is what causes me to be skeptical of the Pentagon's motives in exempting themselves from longstanding acquisition rules. Thank you.

Mr. SHAYS. I thank the gentlemen, and just say to him that the questions he raises, for our guests and also our panelists, we have a lot of hard working members in this committee. We don't have a lot that show up, and so we go over the 5-minutes. General Kadish, you are free to speak, obviously, more than the 5-minutes and answer any questions that you want to address that you hear being raised by people here. Obviously, Mr. Christie and Dr. Stansberry as well.

At this time, I will be calling on my colleague Mr. Schrock.

Mr. SCHROCK. Thank you, Mr. Chairman.

Mr. SHAYS. I would like to introduce you the same way I did my colleague, Mr. Tierney, the most distinguished Mr. Schrock.

Mr. SCHROCK. Thank you very much. Now you've thrown me off c completely. [Laughter.]

I wasn't going to make an opening statement, but I think I'll just make a few comments. I want to associate myself completely with what Mr. Putnam said.

I heard it a few minutes ago when I came in here, we can't choose the threat. If we could, we wouldn't have to be here today and it would make life a whole lot easier. But as far as I'm concerned, the reality of the threat is real, the immediacy of the threat is very, very real. And I hear about tremendous costs and I understand that.

But folks, freedom is not free. And nothing worthwhile is easy or cheap. This is going to be one of those areas as well.

We were never going to get hit. We were never going to get attacked. But folks, we have, and no longer could we take that for granted. We think there are people out there who can't strike us, but I just read, I believe yesterday, that the Chinese now have a missile that could get all the way to Atlanta. That's pretty scary to me.

So I just think we need to understand that if we're going to solve this thing, we've got to spend the money and time and effort it takes to get this system developed. And as far as the comment that DOD is trying to do away with Congress's oversight, I'm not sure there's any evidence of that at all. As far as I'm concerned, the current Secretary of Defense and the staff he's put together, the team he's put together, feel strongly about that and they're trying to do that. So I think that's an unfair assessment.

So again, I thank you all for what you do. I thank you for being here, and I am very anxious to hear what you have to say. Thank you.

Mr. SHAYS. Thank you. Another very active and distinguished member of the committee, Ms. Schakowsky, is recognized.

Ms. SCHAKOWSKY. Thank you for that, Mr. Chairman.

I want to thank you for this hearing and the ranking member, Mr. Kucinich, and also Mr. Tierney for all the work that he has done on this issue.

This isn't the first hearing this committee has had on missile defense. In the past we have discovered deficiencies in just about every facet of the program's development, from testing or lack thereof to acquisition to oversight. Yet each appropriation cycle, Congress spends billions of dollars on this failed system. The only thing we consistently learn from hearings and research on this subject is how much information and accountability is lacking, and how much of a pipe dream this program is.

The Bush administration proposed and Congress recently appropriated nearly $8 billion in funding for this fantasy based device. The United States has already spent the equivalent of $148 billion on research and development since missile defense was first proposed in the 1950's. The latest CBO estimates project that implementation of the Bush administration's missile defense concept will cost as much as $238 billion.

These numbers are astonishing, considering the program's lack of success and even more stunning considering that the administration is, in my view, making it more difficult for Congress to monitor the program.

Today, the GAO will present a report to the subcommittee outlining recommendations for a more knowledge based decisionmaking process at the missile defense agency in order to reduce risks in developing the airborne laser phase of the proposed defense package. I agree with the GAO's recommendations. In fact, many of us are quite familiar with them, because similar recommendations were made in Dr. Phillip Coyle's August 2000 report, which this subcommittee analyzed.

Each time I attend a briefing or read the paper, there is always one very simple point: spending billions of dollars on a system that does not work and will not make us safer is unacceptable.

The Bush administration holds every Government program aimed at social development to the strictest standards of accountability. If this program were in the Department of Housing and Urban Development or the Department of Education, it would be long gone. Yet when it comes to the missile defense system, the Bush administration is trying to evade reasonable and necessary standards of accountability. The Missile Defense Agency has yet to complete even a test and evaluation master plan, program implementation plan or an operational requirements document, as has been stated.

In other words, the administration is spending billions of precious taxpayer dollars on a concept that they haven't even figured out how to test accurately.

Why does the administration cloud oversight and waive accountability for a system that's so expensive? Why does the administration try to hide the development of this system from the Congress and the American people, who pay for it, but consistently tout the success of the program? Why does this Congress continue to appropriate billions of dollars each year with virtually no proof that the system can pass test scenarios that even slightly resemble real life situations, and with no proof that the technologies in question will ever defend our country from missile attack?

These are the questions to which the American people deserve answers. I thank the witnesses for attending this hearing, and I hope that the efforts of the GAO are not simply addressed for the benefit of this hearing. We have a Government Accounting Office for a reason.

I charge the Missile Defense Agency with the responsibility of taking the GAO recommendations seriously and also taking this hearing as a message from the American people that we deserve and demand to know how our money is being spent. In my opinion, if these recommendations are not implemented and if we fail to link funding for this concept to real, clear and convincing scientific facts, further investment in this program will be even more of a waste.

Thank you, Mr. Chairman.

Mr. SHAYS. I thank you very much.

We are blessed to have Mr. Gilman, former chairman of the International Subcommittee and full committee, and has been, frankly, one of the most active members of this committee. It is very appreciated, Mr. Gilman.

Mr. GILMAN. Thank you, Mr. Chairman.

I want to thank you for convening this very important hearing this morning to explore the state of our national missile defense program and its technical feasibility. The missile defense program has undergone a number of significant changes over the last year. Chief among those was the redesignation of the primary agency in charge of the mission, the Ballistic Missile Defense Organization, as the Missile Defense Agency.

In addition to the name change, a major focus of missile defense has also shifted from concentrating on theater against long range issues to develop a more layered defense that emphasizes the various stages of a missile in-flight trajectory.

Last year also saw our Nation formally withdraw from the Anti-Ballistic Treaty of 1972. The Bush administration believed that the Treaty was a cold war relic that had outlived its usefulness. Moreover, the Treaty was threatening to impinge upon near term developments and testing within the U.S. program.

The President made the decision to exercise the withdrawal provisions within the Treaty and the administration made that formal announcement last December. The events of last September have shown that the threat of international terrorism is all too real. This does not, however, render invalid prior concerns about the proliferation of ballistic missile technology. Currently, our Nation has no defense against such a missile being launched deliberately or by any rogue nation or state accidentally by a nation with such technology.

Critics of the missile defense system argue that the chances of such a launch are remote. In response, I content that last year at this time, it seemed equally farfetched that someone could organize a concentrated efforts to fly airplanes into large buildings. Defending against ballistic missiles will never be easy nor inexpensive. But such difficulties in defense should not be any excuse for inaction.

Once again, Mr. Chairman, we thank you for holding this important hearing and we look forward to hearing from today's impressive witnesses. Thank you, Mr. Chairman.

Mr. SHAYS. I thank you very much.

At this time we recognize our last speaker, and also a very valued member of the committee, Representative Allen.

Mr. ALLEN. Thank you, Mr. Chairman. I'll be brief.

Back in Maine, when I try to explain how Congress works, I describe this subcommittee as operating the way people think Congress works, that is, bringing people in and having a full blown discussion so we can have good information and an exchange of views. I wish more committees and subcommittees operated this way.

I'll be very brief. I have been, as people know, concerned about the transformation of the missile defense program into one that where it is difficult to get costs, difficult to evaluate progress, difficult to understand what's going on, because all the missions have been lumped together.

I will save my comments and questions primarily, but I did want to react to one comment by Congressman Schrock. It's always been my understanding that the missile defense program that the United States is developing is not designed to shoot down missiles from either China or Russia or Britain or France, for that matter, and that it is designed to deal only with the threat from rogue states. If we were going to deal with the Chinese and Russian threat, we would probably have to double the current budget.

If I'm wrong in that assumption, I would appreciate it if the witnesses would react to that today. Because it's one example of, I would say, the importance that we keep focused on what it is we are trying to do, what the nature of the threat is, and making sure that we are designing and developing a system that is responsive to the particular threat that is perceived out there. That's the ground of my concern for the direction the agency is headed in now.

Mr. Chairman, thank you again for holding this hearing. I'm pleased to be here.

Mr. SHAYS. I thank the gentleman.

Before swearing in the witnesses, I just want to thank them for their patience in listening to the comments made by Members. I think it will help everyone respond, because we really know what's on the table. I thank all of our Members for placing on the table their concerns and comments of support.

But I do want to particularly say, it's been the first hearing this subcommittee has had since Mr. Gilman announced that he would not be running again. He is the first Member I ever met, he has been one of the most gracious Members to serve this chamber, one of the most knowledgeable. And when some of the newer Members are not attending this subcommittee, he attends and participates as if he were newly elected. I just appreciate his graciousness, his intelligence, his lack of partisanship. You are to me a real hero and a person I would love to emulate if I could.

Mr. GILMAN. Thank you, Mr. Chairman, for your kind words.

Mr. SHAYS. And I know they're shared by all the Members. So we won't spend more time talking about you now, but we will later.

I would like to recognize our first panel. We have General Ronald T. Kadish, U.S. Air Force, Director, Missile Defense Agency, Department of Defense. We also have testifying Mr. Thomas Christie, Director, Office of Test and Evaluation at DOD. And we have Kent G. Stansberry, Deputy Director, Missile Warfare, Office of the Secretary of Defense, Department of Defense.

Gentlemen, as you know, we swear in our witnesses. I invite you to stand and we will swear you in.

[Witnesses sworn.]

Mr. SHAYS. Note for the record our witnesses have responded in the affirmative.

If I could just do the housekeeping, that enables us to begin. I ask unanimous consent that all members of the subcommittee be permitted to place an opening statement in the record and that the record remain open for 3 days for that purposes. Without objection, so ordered.

I ask further unanimous consent that all witnesses be permitted to include their written statement in the record and without objection, so ordered.

Gentlemen, what we will do with the clock, we do the 5-minutes, we roll it over. Frankly, given all the comments that were made by the Members here, if you need to go a little longer, so be it. This is a very important hearing to us. We don't want you to feel that you're not able to put everything you need to put on the record on the record.

We'll start with you, General.

STATEMENTS OF LIEUTENANT GENERAL RONALD T. KADISH, U.S. AIR FORCE, DIRECTOR, MISSILE DEFENSE AGENCY, DEPARTMENT OF DEFENSE; THOMAS P. CHRISTIE, DIRECTOR, OPERATIONAL TEST AND EVALUATION, DEPARTMENT OF DEFENSE; AND KENT G. STANSBERRY, DEPUTY DIRECTOR, MISSILE WARFARE FOR THE UNDER SECRETARY OF DEFENSE (ACQUISITION, TECHNOLOGY AND LOGISTICS)

General KADISH. Good morning, Mr. Chairman and members of the committee. Thanks for inviting me to establish and testify on behalf of the Missile Defense Agency and reorganization of that program.

I want to address briefly just three aspects of our program: management, requirements and testing. Our goal in developing missile defense is to be effective against all ranges of ballistic missiles. It is a national decision as to where and when we deploy our capabilities. But in developing them, we face the complex task of integrating many elements, because the flight physics involved, the variety of missile speeds, trajectories and environments through which the missiles travel preclude our having one defensive technology or approach that can do it all.

In reforming our management approach, we looked at what made current and past complex national programs successful and took good counsel of what our internal and external critics had to say. Our review has led to a management approach that meets the challenges of the unprecedented technology needed for effective missile defense. I believe this management approach is sound and will lead to successful development and deployment of such an effective missile defense.

Now, why did we change our approach? There are two major reasons. First, to reduce the cycle time for making key decisions. Our revised structure provides for more direct, focused, frequent and comprehensive decisionmaking. It is designed to attack head-on the tough problems of systems-of-systems integration, which is so key to the success of such a complex undertaking as missile defense.

The second reason we have changed our processes is that the existing Departmental procedures were designed to satisfy the needs of single service acquisition. Missile defense is different. In missile defense, we have three military departments, the joint staff and the combatant commanders all deeply involved. That is why the Missile Defense Agency was created in the first place, to pull all these strands together, regardless of whether the basing mode of any single element was on the ground, at sea, in the air, or in space, to create a single layered integrated missile defense system.

Managing the missile defense program under these circumstances requires a new approach, one that builds on the best practices we know how to implement. I believe the Secretary of Defense's direction provides a foundation for that approach, and we are making good progress in executing that direction.

Our approach to the problems of the traditional requirements process is capability based acquisition. Capability based acquisition simply allows us to design flexible systems that can accommodate new technologies and concepts, while fielding demonstrated capabilities more quickly as they mature.

This ability to be flexible during development allows us to reduce acquisition cycle time, schedule risks and costs risks. In this way we can find the right balance between what is needed and what is possible at any point in the acquisition path.

The significant advantage is that capabilities based acquisition promotes a potential early deployment of missile defense capability that has military utility. Even if that capability is limited, it fills a serious gap in our current national security posture. That is because today, our Nation has no defense against long range missiles and only very limited defense against short range missiles.

Now, as I look back, it has been a very busy year, highlighted by a series of important events just a month ago. In the space of 3 days, we had a successful intercept by a standard missile three interceptor launched from an Aegis cruiser, the United States formally withdrew from the ABM Treaty, and we broke ground on the start of an expanded ballistic missile system test bed at Fort Greeley, AK. These events underscore the fact that we are truly at a crossroads in the development of missile defense. Our pace has picked up, and it is important that we sustain our momentum.

A robust and progressive test and evaluation program is absolutely necessary to make sure our technologies are mature enough. Our tests have shown that we have the technology in hand to hit a missile. We have done it in space and in the atmosphere from land and from the sea. Now we have to move to the next two important stages of testing, to show that we can do that reliably and to show that we can do it against countermeasures.

The most visible evidence of our progress is success in flight tests. At this point, if you will allow me, I would like to show a video score card of the tests that we have conducted in the last 2 years. If you could put the video on, please.

[Video shown.]

General KADISH. The ballistic missile defense system has some major tests in the atmosphere and in outer space. We are looking at layered defenses.

What that means is that we have blue space, midcourse and terminal defense systems that handle long range, medium range and short range threats. This is designed to make it much more complicated for an adversary to punch through a layered defense.

Now, Patriot–3 is a terminal defense system against short range missile. The intercept occurs in the atmosphere. What you will see is a Patriot–3 launch. You see it maneuver into position to very accurately hit the incoming missile. As it approaches the missile, you will see the rockets firing very accurately. No explosives on that missile, it is pure hit to kill energy.

Now we move up into the atmosphere, into the trans-atmosphere and the edge of outer space with the THAAD system. This is a terminal system against longer range missiles and medium range missiles. Here is the launch of the target at White Sands Missile Range and two views of the interceptor launching. The missile actually has to do an energy dissipation maneuver to stay in range.

This interception occurred in outer space. You can see the maneuvering vehicle, kill vehicle heading toward the target and a slow motion intercept. And another view of seeing that real time. And another perspective of the intercept from the ground.

The next thing you will see is what the seeker on that interceptor actually saw microseconds prior to intercept. It gives you an idea of the accuracy with which these early tests have proven we can actually do hit to kill with.

Now we have a booster test. This is a booster test of the longer range system, Midcourse, the ground based missile defense system. Unfortunately, this test was not successful. It was not an intercept defense test, it was a test of the booster that we intend to use. For those of us in the program, this was very disconcerting to watch; 15 seconds into the flight, it encountered a failure and basically self-destructed.

We know how to fix this problem and we intend to fix it.

We had, however, some remarkable success in our ground based program, especially over the past year. I would like to show two videos of this particular set of testing. The first one, they are basically all the same. Launched out at Vandenberg, headed toward the South Pacific. It will rise to meet that incoming warhead somewhere in the neighborhood of 240 kilometers in altitude in outer space.

This is what the seeker saw as it came in at very high speed, closing velocity 15,000 miles per hour. We actually very accurately destroyed that target. This is another view of it coming up. This one was done March 15th of this year.

Again, same test, to make sure we can do it repeatedly. Again the target launched out of Vandenberg, and the interceptor at approximately 4,500 miles away. And you'll see a different view of both the intercept and the interceptor prototype that we're using.

These tests are extremely complex. We have many sensors and we have much data gathering going on. This is an infrared view of the actual intercept. The next view you will see both a visual and radar tracking data on that intercept.

The last view I would have to show you is what we did from the sea with a standard missile against midcourse intercepts of medium range ballistic missiles. This test occurred at the Pacific Missile Test Range located in Hawaii, with a launch of a target off of Hawaii, about 500 kilometers downrange. This is a view of the cruiser and the launch of the standard missile three to intercept that target.

Here is another view, it's very fast. This missile intercept took place in outer space. This is one view of it with a chase plane. And this is another view from the seeker, what the seeker actually saw microseconds prior to impact.

Now, we learn from our failures. And we advance with our successes. If every test were a success, the envelope probably is not being pushed hard enough. On the other hand, if we have too many failures, we're obviously not doing something right. Our goal is success, and we are achieving that as our testing base and complexity pick up.

Over the past year, in tests against long range missile targets by the ground based midcourse defense system, the interceptor record is three for three. But the full record is four for six. Early tests focused on technology development demonstration and integration. Now we're advancing step by step to the operation of more realistic and more demanding scenarios.

Our sea based interceptor tests against medium range target missiles has gotten off to a good start. We are two for two in intercepts. These two successes move us along to the next stage of development, to capitalize on the advantages they have brought to the schedule.

Our Patriot Advanced Capability–3, or PAC–3, intercept record against shorter range ballistic missile targets is 7 for 10 and 2 for 4 over the past year in tougher operational tests. The mixed results from these tests have shown we still have a little bit more work to do.

Our momentum is continuing. To mitigate the risk of flight test failures, we are aggressively pursuing a wide range of ground tests across the program. Over the next 6 months, we have some additional 15 significant ground tests scheduled. Additionally, we have some 20 flight tests scheduled, including several phenomenology flight tests.

As we move into a new era of testing, one no longer constrained by the ABM Treaty, we can now test new concepts of missile defense that may lead to a new and better way to accomplish the BMD mission. This will of course change what and how we test, so as to find the best role and mix of components and elements to provide the best overall system.

Mr. Chairman, the missile defense program is at a crossroads. With the continued support of Congress, we can keep up with the aggressive pace of the momentum of our development effort and set the stage for the successful deployment of missile defenses against all ranges of threats.

I apologize for running over, but I thought this was a complete view of our program.

[The prepared statement of General Kadish follows:]

Lt Gen Ronald T. Kadish, USAF
Director, Missile Defense Agency
On the Missile Defense Program to the
Subcommittee on National Security,
Veterans Affairs, and International Relations;
House Committee on Government Reform

July 16, 2002

Good Morning, Mr. Chairman, Members of the Committee. Thank you for inviting me to testify on the establishment of the Missile Defense Agency and reorganization of the Missile Defense Program.

The changes in the program over the past year and a half have been driven by the worrisome trends in proliferation of ballistic missiles and weapons of mass destruction. There are those who wish our nation ill, as September 11 brought forcefully home. Some may be developing suitcase bombs and planning terrorist attacks, but others undeniably have active programs to build missiles of increasing range and lethality. Security is not an either-or proposition.

Earlier this year, Secretary of Defense Rumsfeld stated the issue clearly: "We're at a moment where we no longer have the margin for error [that we] … had decades ago where our weapons were relatively short range and where the warheads were relatively modest.[1]" We can expect to be surprised again, and the consequences could be grave.

Furthermore, as he told the NATO Defense Ministers last month, "Rogue states capable of delivering weapons of mass destruction to Western Capitals could make building future coalitions against aggression difficult, if not impossible." That is why the President directed the development of a system that could defend our allies and friends, as well as U.S. territories and deployed forces.

<u>Program Priorities, Direction, and Structure</u>

To carry out the President's direction, the Secretary gave us guidance on January 2nd of this year, when he redesignated the Ballistic Missile Defense

[1] Interview with [Britain's] *The Daily Telegraph*, 25 Feb 02.

Organization as the Missile Defense Agency and changed the responsibilities and authorities of the Director.

The Secretary gave the Agency, and me as the Director, new priorities and direction, and expanded responsibilities and authority to execute the missile defense program. Some of these new authorities differ from traditional Departmental processes, but all of them are within the Secretary's existing statutory powers. We believe that the changes we have instituted will allow us to improve management and execution of the missile defense program. We will continue to be responsible stewards of the resources entrusted to us. Our accountability to you and the American people will be just as assured as it has been in the past.

The Secretary gave the Department four top priorities for missile defense. They are:

- First, to defend the United States, deployed forces, allies, and friends;
- Second, to employ a Ballistic Missile Defense (BMD) System that layers defenses to intercept missiles in all phases of their flight (i.e., boost, midcourse, and terminal) against all ranges of threats;
- Third, to enable the Services to field elements of the overall BMD System as soon as practicable; and
- Fourth, to develop and test technologies, use prototypes, and test assets to provide early capability, if necessary, and improve the effectiveness of deployed capability by inserting new technologies as they become available or when the threat warrants an accelerated capability.

The Secretary also provided his specific objectives for the Missile Defense Program, to:

- Establish a single program to develop an integrated system under a newly titled Missile Defense Agency (MDA);
- Assign the best and brightest people to this work;
- Apply a capability-based requirements process for missile defense; and
- Direct the MDA to develop the missile defense system and baseline the capability and configuration of its elements and the Military Departments to procure and provide for operation and support.

In response to this direction, we are capitalizing on developmental progress already achieved, while at the same time pursuing new development opportunities previously foreclosed by the ABM Treaty. In so doing, we are moving from being element-centric to system-centric. This is a long-term approach to a long-term threat.

Let me clarify some of the expressions we use. When we speak of one BMD System, we refer to the operational integration of all missile defense elements, including sensors, weapons, and battle management/command and control capabilities, regardless of which Service operates them. We speak in terms of three defense segments that categorize the capabilities to intercept a hostile missile during each phase of its flight: a Boost Defense Segment, a Midcourse Defense Segment, and a Terminal Defense Segment. These are the layers through which any hostile missile must pass before it reaches its target and within which we are developing integrated defenses.

In programmatic terms, we no longer speak of national or theater missile defense. Operationally, the terms can take on different meanings depending on where you live. The distinction between them made sense a decade ago, when we faced the stark difference between a Soviet ICBM threat and an Iraqi Scud. Now it no longer does. The same North Korean missile aimed at Japan could be a national threat to our ally, but a theater threat to us—unless it were retargeted toward the United States, in which case it would become national again. Furthermore, at some point in time, a short-range missile could threaten our homeland just as well as an ICBM could, if, for example, it were launched from the sea off our coast.

Our goal is to be effective against all ranges of ballistic missiles. It is a national decision as to where and when we deploy our capabilities. We face the complex task of integrating many elements, because flight physics—the variety of missile speeds, trajectories, and the environments through which missiles travel—precludes our having one defensive technology or approach that can do it all.

Acquisition Management for a Focused Missile Defense Program

In reforming our management approach, we looked at what made complex national programs successful and took good counsel of our internal and external critics. I believe our management approach is sound and will lead to a successful development and deployment of an effective missile defense.

I continue to report directly to the Under Secretary of Defense (Acquisition, Technology and Logistics). The Senior Executive Council, or SEC, chaired by the

Deputy Secretary, provides executive oversight of the program. Permanent members are the Service Secretaries and the Under Secretary (AT&L). Other Department officials will be included as needed, depending on the subject at hand.

The SEC conducts periodic formal and informal reviews of the program. I have met with the council six times since last summer. And I will report to it again later this summer to present our program and budget proposals for next year. Reviews include such topics as program plans, management approaches, test performance, system architecture, technological alternatives, basing options, and threat. The SEC provides guidance regarding policy, planning, and programming; makes the decisions as to whether to stop, start, slow, or accelerate efforts; and approves recommendations on fielding elements of the system. This group demands high standards of accountability. You saw some of this last winter with the Department's decisions to cancel the Navy Area Defense program and restructure SBIRS-Low.

Additionally, the Department has created a standing Missile Defense Support Group, or MDSG, the Chairman of which reports directly to the Under Secretary (AT&L). The group has representatives from each of the 13 separate Departmental entities with a stake in our program, and its members are all senior and experienced in missile defense. The MDSG provides advice both to the Under Secretary and to me, as well as input to the SEC. While we are still working through normal organizational startup issues with the MDSG, we are making good progress. It has met virtually every two weeks since it was stood up over four months ago.

Why these changes? There are two major reasons. The first is to reduce the cycle time for making key decisions. The structure provides more direct, focused, frequent, and comprehensive decision making and is designed to attack head-on the tough problems of complex "system-of-systems" integration that are key to the success of such a complex undertaking as missile defense.

If we went through the normal Departmental acquisition processes, the reviews would be episodic, subject to individual program events or milestones, and there could be years between these events. Our new procedures provide for more frequent and more comprehensive review of the missile defense program. We can respond more effectively to changing conditions and emerging events. Our current management procedures provide for more internal accountability and for decision making at a more rapid pace than we have had in the past.

The second reason we have changed our processes is that existing Departmental procedures were designed to satisfy the needs of single Service acquisition. Very seldom have Service boundaries been crossed. Even when the Department deals with very complex programs, such as the F-22, the DD-X, or the Comanche attack helicopter, ultimately we are looking at one Service to develop and operate the system. Our DoD acquisition procedures have been designed over time to manage development and fielding under that one Service.

Missile defense is different. In missile defense, we have three Military Departments, the Joint Staff, and the combatant commanders all deeply involved. Service boundaries are crossed from the start of development to provide for complex integration within the BMD System. Even at the operational level, no one Service will operate integrated, layered missile defenses. That is why the Missile Defense Agency was created in the first place, to pull all these strands together, regardless of whether the basing mode of any single element was on the ground, at sea, in the air, or in space.

Managing the missile defense program under these circumstances requires a new approach—one that builds on best practices to meet the challenges of a complex and well-focused program that can achieve the multi-Service "system-of-system" integration required of a layered BMD System. I believe the Secretary of Defense's direction provides a foundation for this approach and we are making good progress in executing that direction.

Setting Requirements for the BMD System

In order to exploit the unprecedented nature of missile defense technology, we are using a capability-based requirements approach to acquisition, rather than the traditional threat-based requirements approach. Some have interpreted this as doing away with requirements. That is not the case. We are not doing away with requirements. Let me explain.

Instead of developing a system to respond to a narrowly defined threat from a known adversary, we find we cannot know with confidence what specific adversary might pose what specific missile threat or when. Hence, as a starting point, we are looking at the broader range of capabilities an adversary might have in a given timeframe and then developing defensive capabilities based on technology that will evolve over time.

The traditional acquisition process starts with specific military requirements generated by the user that become formalized in the Operational Requirements

Document, or ORD. This traditional ORD approach has generally served our nation well, especially in procurements involving well-known technologies, proven systems, sizeable production runs, established operational experience, and single-Service acquisition. None of these, however, yet exist in missile defense—and may never.

For missile defense, the strength of the traditional requirements generation process can also be its weakness. The procedures are rigorous, but the way the rigor was applied resulted in a lack of flexibility for dealing with the unprecedented technology development and integration challenges we face. When the traditional requirements process is applied to missile defense, we have a high risk of starting formal programs before the technology is mature enough. In addition, requirements defined in ORDs are typically set many years before actual system deployment. Therefore we run the risk that the technology eventually deployed could be both a generation or two behind the industrial leading edge and represent less than optimum capability against a threat that changes over time from proliferating states.

Furthermore, at the moment, we do not yet know which technical approaches will work best for missile defense. For example, five years ago, we could not have foreseen, let alone written down, all the uses that define today's Internet. It would not be prudent to lock in our development path now with a "grand design" and find out some years down the road that we have weakness in the system. This could come about because of an unexpected technical obstacle; because we failed to exploit some practical technical innovation uncovered earlier; or because of some new development in the threat.

Yet we always face the risk of being surprised by changes in the threat. Missile defense has perhaps more uncertainties in this regard than many other mission areas. We do not want to alter our baseline every time we discover a change in the threat. Such changes could ripple through the program to cause significant delay and cost. So instead of designing against a point threat, we are setting a wider range of boundaries for adversarial capabilities over time in defining our own missile defense capabilities.

Our approach to the problems of the traditional requirements process is capability-based acquisition. Capability-based acquisition allows us to design flexible systems that can accommodate new technologies and concepts, while fielding demonstrated capabilities more quickly, as they mature. This ability to be flexible during development allows us to reduce acquisition cycle time, schedule risk, and cost risk.

While we are moving away from some of the rigidities of the past, we are not abandoning rigor in development. In my opinion, far from it. A capability-based approach provides for significant discipline. It is just guided by different mileposts. Instead of the traditional process where users define the requirement in great detail, then, subsequently, developers translate the requirements into specifications, we intend to do both at the same time. In so doing, we can accrue the same advantages that cutting-edge development in the commercial world already enjoys.

We have already taken a step in that direction. I have approved the initial Technical Objectives and Goals document, or TOG, that defines the capabilities to which the BMD System should be designed. This document will be the starting point for the development of the BMD System and its elements.

We are bringing together users and developers under our Agency's lead—the warfighters, the Services, the testers, and industry. Together, all of them will have a continuous and constructive role in establishing the mission requirements for missile defenses. The warfighters will not disengage after setting the requirements at the outset, and industry will not be brought in at the last moment—they will both be present from the start. This partnership will be continuous and remain vital throughout the whole development process.

In this way, they can find the right balance between what is needed and what is possible at any point in the acquisition path. The focused, continual interaction between user and developer will allow us to make timely capability trades, explore a broader range of options, and upgrade our capabilities to stay ahead of the threat and to keep them current.

When we decide to move to production, a description of operational capabilities is drawn up. It is one with which the Services and the oversight community are fully onboard, and it is based on progress proven to that point. This document will govern procurement just as would the traditionally derived ORD.

The significant advantage of this acquisition approach is that it permits the early deployment of a missile defense capability that has military utility. Even if that capability is limited, it fills a serious gap in our current national security posture. That is because, today, our nation has no defense against long-range ballistic missiles, and only a very limited defense against shorter-range missiles.

This acquisition approach is not really new. We have learned from past successful programs—our nation has used it in undertaking previously unprecedented technological endeavors. For example, we used this acquisition approach for making trade-offs and accelerating the schedule to develop the Polaris submarine-launched ballistic missile and the SR-71 reconnaissance aircraft. And we are certainly familiar with upgrading systems over time. The B-52s that flew over Afghanistan last fall were far different aircraft than those that first rolled off the production lines five decades ago.

To recap, this approach to acquisition involves a change in the traditional way we derive, define, and deal with requirements, not do away with them. Capability-based acquisition allows us to design flexible systems that incorporate new technologies and concepts, while fielding demonstrated capabilities more quickly, as they mature.

Conducting and Assessing Tests

Fundamentally, the core of our testing philosophy remains the same—we are following a step-by-step, disciplined approach to testing that allows us to build on lessons learned at each stage of development. It is a walk-before-you-run approach. Nevertheless, there will be some significant changes in emphasis in how we go about that testing. Systems engineering, ground testing, flight test planning, and systems integration testing are all receiving renewed attention and significant budgetary resources. Our new test bed concept in the Pacific reflects this new emphasis.

Certainly, testing of missile defenses will become significantly more complex, and not just because we will be progressing from developmental to operational testing. In addition to the autonomous operation of the elements, testing must consider the integration of the entire system, and multiple potential system architectures have to be evaluated. Furthermore, testing plans must be flexible enough to adjust to tradeoffs that could be made in midstream, thus requiring a reevaluation of testing plans and goals. And the expanded testing infrastructure, especially the BMD System Test Bed, will allow us to test portions of the operational envelope that we could only simulate before.

A major change will involve the object of testing—a block capability, not a final product. Testing will continue along the spiral development path, becoming increasingly rigorous throughout development. Testing will not be concentrated just on the initial validation of requirements, with episodic follow-on for improvement.

Another change is the goal of the testing— to demonstrate a useful military capability, rather than to satisfy a fixed set of specifications. The testers will be fully engaged along with users and developers in defining and developing the end product, with the goal of delivering a capability where it is needed as soon as possible. This means that, unlike the traditional system, the testers help write the description of operational capability governing the production of block capability, in close cooperation with the development process.

This close involvement of testers is essential if we are to be able to deploy a capability early. It also requires independent review of testing procedures and results. We have provided for this as follows.

Our revised acquisition process now specifies three broad phases: Development, Transition, and Procurement and Operations. As Director, MDA, I have oversight and responsibility for managing the first two phases, Development and Transition. The SEC, upon my recommendation, approves the advance from Development to the Transition phase. The Services manage Procurement and Operations.

The Director, Operational Test and Evaluation (OT&E), plays a continuing role in each of the phases and is in a position to ensure an objective assessment of testing results. Mr. Christie will go into greater detail as to his role in oversight and reporting.

During the Development Phase, as Director, MDA and the lead developer for missile defense, I retain the traditional responsibility for developmental testing, just as I did as Director, Ballistic Missile Defense Organization. In the new context, developmental testing will help us to characterize a capability and assess its military utility, before making recommendations for fielding such a capability.

The most visible evidence of progress in the development and integration of technologies in missile defenses is success in flight tests. Yet these represent only the tip of the iceberg in testing. Modeling and simulations and ground testing undergird all tests at almost every level of technological maturity. Testing starts in the laboratory or machine shop, and continues up through ground tests, flight tests, intercept flight tests, and ultimately must be assessed at the system-wide level of effectiveness.

Because flight tests can be very expensive—on the order of $100 million per GMD integrated flight test, for example—we make every effort to get the most out of each one, so planning is detailed. Our risk mitigation strategy to avoid test failures at this level is based on thorough ground testing, a key but often unnoticed part of our program. We are aggressively pursuing a wide range of ground tests across the program with very positive results. Deliberate ground tests sow the seeds of success in our future flight tests. We have some 15 significant ground tests coming up over the next six months.

A second strategy to minimize test failures is to avoid overloading them with too many objectives. Should a failure occur, we want to be able to trace its root causes so as not to repeat it. It is a disciplined, rigorous, and time-tested approach that has served the nation well.

Testing during development is always a balancing act. We learn from failures, and we advance with success. If every test were a success, the envelope is probably not being pushed hard enough. On the other hand, if we have too many failures, we are obviously not doing something right. Our goal is success, and we are achieving that as our testing pace picks up and the complexity of our testing scenarios increases.

Testing Progress

In contrast to previous years, we have shown that the technology of hitting a missile in space or the atmosphere works. Now we need to show that it works reliably and that it works against countermeasures. And our program is on track to do just that.

Over the past year, in tests against long-range missile targets by the Ground-based Midcourse Defense, the intercept record is 3-for-3, with the full record at 4-for-6. Early tests focused on technology development, demonstration, and integration. Now we are advancing, step-by-step, to be more operationally realistic in more demanding scenarios.

Our sea-based interceptor tests have gotten off to a good start. The first two tests of the Sea-based Midcourse Defense (SMD) against a medium-range target missile, one last December and the second last month, both resulted in direct hits in space. With these two tests, we have fulfilled our initial basic objectives and are moving on to the next stage of development. We do not have to prolong our initial test series, and we can capitalize on the advantages these early successes have brought to the schedule. We can do that because we have been able to get our decision cycle time down.

Against shorter-range ballistic missile targets, the Patriot Advanced Capability-3, or PAC-3, intercept record is 7-for-10, and 2-for-4 over the past year in the tougher operational tests. The mixed results from these tests have shown we still have some more work to do.

Less Restricted Testing

The President's decision to withdraw from the ABM Treaty has had a major impact on our missile defense program—in design, development, operation, and, of course, in testing. It has opened opportunities for new approaches to create a more effective layered defense, and testing can take advantage of restrictions now no longer in place. Other treaties, of course, are still in effect that impact on the way we can test.

Testing of ABM components (to support the development of a BMD System to defeat long-range or "strategic" ballistic missiles) is no longer limited to the two designated ABM test ranges: the Reagan Test Site anchored at Kwajalein, and the White Sands Missile Range in New Mexico. The planned expanded BMD System Test Bed, extending across more of the Pacific, will allow testing of BMD System elements and components along new geometries and in new combinations.

Additionally, testing and deployment of components as part of a missile defense system to defeat long-range ballistic missiles can now be done in any basing mode (i.e., sea-based, space-based, air-based, and mobile land-based), all of which were prohibited by the ABM Treaty. Similarly, the concurrent testing of ABM and non-ABM components is no longer prohibited.

The immediate benefit of these changes is to allow us to test the Aegis radar in conjunction with our Ground-based Midcourse Defense integrated flight tests, the next of which is scheduled for later in the summer. Upcoming tests will involve having the shipboard SPY-1 radar off the coast of California track the outbound missile in its boost and ascent phase from Vandenberg AFB, and to have another in the vicinity of Kwajalein track the interceptor. At the same time, we plan to use the Multiple Object Tracking Radar at Vandenberg to track the outbound target. Initially, these tests will evaluate simply how well they can perform the tracking function in a more stressful environment. Later tests will evaluate how well they can provide actual tracking data for use by the integrated system.

Other tests will involve incorporating the THAAD radar with GMD tests, and involving the Ground-Based Radar prototype (the GBR-P), formerly used only in ABM work, to support PAC-3 missile tests at Kwajalein.

81-892 D-2

These preliminary tests will help validate the effectiveness of these components in roles for which they were not initially designed, but in which they could prove most useful. The tests will inform decisions regarding modifications to make these components better suited for integration into the BMD System in later blocks of capability.

This is perhaps the biggest influence of less restricted testing—that we can test new concepts of missile defense that are less constrained and may lead to new and better ways to accomplish the BMD mission. This will, of course, change what and how we test so as to find the best role and mix of components and elements to provide the best overall system.

Closing

Mr. Chairman, we have modified our approach to the acquisition of missile defenses because of the need to respond more rapidly and effectively to the changes in the ballistic missile threat to our nation's interests. The Department has restructured the program for missile defense development so that we can substantially reduce our decision cycle time to handle a very complex set of technical and operational challenges. In doing so, we have taken good counsel of what made some of the challenging programs of the past successful, and we have heeded the sound recommendations of our critics, both internal and external.

The missile defense program is at a crossroads. With the continued support of the Congress, we can make the right choices to keep up the aggressive pace and momentum of our development effort and to set the stage for the successful deployment of missile defenses against all ranges of threats.

GLOSSARY

ABM	Anti-Ballistic Missile
AT&L	Acquisition, Technology and Logistics
BMD	Ballistic Missile Defense
ICBM	Inter-Continental Ballistic Missile
GBR-P	Ground-Based Radar Prototype
GMD	Ground-based Midcourse Defense
MDA	Missile Defense Agency
MDSG	Missile Defense Support Group
NATO	North Atlantic Treaty Organization
ORD	Operational Requirements Document
OT&E	Operational Test and Evaluation
PAC-3	Patriot Advanced Capability 3
SEC	Senior Executive Council
SBIRS Low	Space-Based InfraRed System Low
SMD	Sea-based Midcourse Defense
THAAD	Theater High Altitude Area Defense
TOG	Technical Objectives and Goals

Mr. SHAYS. No apology necessary. We would have been disappointed if you hadn't run over.

At this time we will go to you, Mr. Christie.

Mr. CHRISTIE. I also thank you, Mr. Chairman and members of the committee, for this opportunity to appear before you and discuss the involvement of my office, that of the Director of Operational Testing and Evaluation, in this missile defense testing.

As General Kadish has described on several occasions, the ballistic missile defense system will be developed and acquired using a new strategy that incorporates a phased introduction of missile defense capabilities based on evolving technology. Capability based requirements, if that's what you want to call them, are replacing traditional operational requirements and research and development will focus on maturing technologies with operational potential before transitioning them to formal acquisition programs.

My responsibilities in this process include monitoring the demonstration of those critical technologies, providing my senior leadership with advice on missile defense agency goals and plans, and assessing the adequacy and sufficiency of the ballistic missile defense system test program. Traditional operational test oversight will apply once these capabilities have transitioned to the services for acquisition.

Statute requires that I review and approve test plans for both operational and live fire testing, as well as oversee and evaluate these test programs. Live fire testing in particular requires early involvement in the systems development phase. Aside from live fire testing, statute limits my role in developmental testing to that of an advisor. In response to the fiscal year 2002 Defense Authorization Act, I will provide a report to Congress each year by February 15th detailing my assessment of the Missile Defense Agency's testing.

There is proposed fiscal year 2003 language which would further require that I provide operational assessments for ground based missile midcourse defense, sea based missile defense, theater high altitude aerial defense and air boat based booster systems.

As I recently testified before the House Armed Services Subcommittees on Military Research and Development and Military Procurement, I will satisfy the intent of that language under the current reporting requirements. Operational assessments of systems in development are essentially technology assessments with an eye toward operational suitability and effectiveness.

Through my role as the monitor of technology demonstrations and advisor to the director of the Missile Defense Agency, and other senior Department officials, I will provide assessments of evolving technologies on an ongoing basis. My staff and I are being afforded access to important programmatic plans. For example, we were recently briefed in some detail on the ground midcourse defense system test bed and other missile defense range and resource plans and requirements.

While it is premature to assess the adequacy of individual initiatives, the Missile Defense Agency plans certainly seem at this point to be headed in the right direction. When fully implemented, the test bed will mitigate many of the existing test limitations and restrictions that prevented operationally realistic flight testing identified during planning for the former national missile defense program. Testing will gradually increase in complexity and tactical realism as this test bed matures. Furthermore, planning for preflight tests, modeling and simulation and other specialty ground testing will continue to mitigate flight test risks.

The Missile Defense Agency is also taking the initiative to create a combined task force approach for ground midcourse defense that has proven successful in other military systems development. The integration of contractor and Government developmental and independent operational testers has created a team that is working together now to plan, conduct and assess the results of major ground midcourse defense testing. This approach permits earlier operational testers' involvement while still focusing on the examination of technical and design issues, as the system matures through research and development.

I will rely on this team as the primary information source in preparing my assessment, which will support senior executive counsel and defense acquisition board transition and deployment decisions.

In order to effectively and efficiently meet these new as well as traditional reporting responsibilities, I do require extensive access into the Missile Defense Agency's activities. This will be accomplished at three levels. First, through the missile defense support group that Secretary Aldridge has established. Second, through regular communications with the Missile Defense Agency. And third, through my staff's direct involvement at the program and element level.

General Kadish and I both testified as recently as 2 weeks ago that DOT&E is being provided the necessary access to Missile Defense Agency programs. For example, in the Patriot Advanced Capability–3 program, which is in its operational testing phase, we are involved on a daily basis. Due to the reduced level of activity of the other programs during the restructuring that took place late last year and earlier this year, we are just now increasing our involvement with them. We are fast approaching the level of access and involvement that I deem necessary. General Kadish and I are both committed to making it the best way of working that it can be.

Mr. Chairman, I thank you for the opportunity to make these remarks.

[The prepared statement of Mr. Christie follows:]

Opening Statement by

Thomas P. Christie

Director, Operational Test and Evaluation

Before The

House Committee on Government Reform

Subcommittee on National Security, Veterans Affairs,

and International Relations

National Missile Defense: A New Organization, Evolutionary Technologies, and Unrestricted Testing

July 16, 2002

For Official Use Only
Until Release by the
Committee on Government Reform
U.S. House of Representatives
Reform – July 16, 2002

Good morning, Chairman Shays, members of the committee. I want to thank you for the opportunity to appear before you and discuss the involvement of the office of the Director of Operational Test and Evaluation (DOT&E) in Ballistic Missile Defense testing.

As Lieutenant General Kadish has described on several occasions, the Ballistic Missile Defense System will be developed and acquired using a new strategy that incorporates a phased introduction of missile defense capability, based on evolving technology. Capability based requirements are replacing traditional operational requirements and research and development will focus on maturing technologies with operational potential before transitioning them to formal acquisition programs. My responsibilities in this process include monitoring the demonstration of these critical technologies, providing senior leadership with advice on Missile Defense Agency (MDA) goals and plans, and assessing the adequacy and sufficiency of the Ballistic Missile Defense System test program. Traditional operational test oversight will apply once capabilities are transitioned to the Services for acquisition.

Statute requires that I review and approve test plans for both operational and live fire testing, as well as oversee and evaluate these test programs. Live fire testing in particular requires early involvement in the system development phase. Aside from live fire testing, statute limits my role in developmental testing to that of an advisor. In response to the FY02 Defense Authorization Act, I will provide a report to Congress each year by the 15th of February, which details my assessment of MDA testing. Proposed FY03 language would further require that I provide operational assessments for Ground-Based Midcourse Defense (GMD), Sea-Based Midcourse Defense, Theater High Altitude Area Defense, and Air-Based Boost elements. As I recently testified before the House Armed Services Subcommittees on Military Research and Development and Military Procurement, I will satisfy the intent of that language under current reporting requirements. Operational assessments on systems in development are essentially technology assessments with an eye toward operational suitability and effectiveness. Through my role as a monitor of technology demonstrations and advisor to the Director, MDA and senior Department officials, I will provide assessments of evolving technologies on an ongoing basis.

My staff and I are being afforded access to important programmatic plans. For example, we were recently briefed in detail on the GMD test bed and other missile defense range and resource plans and requirements. While it is premature to assess the adequacy of individual initiatives, the MDA plans certainly seem to be headed in the right direction. When fully implemented, the test bed will mitigate many of the existing test limitations and restrictions that prevented operationally realistic flight-testing identified during planning for the former National Missile Defense program. Testing will gradually increase in complexity and tactical realism as the test bed matures. Furthermore, planning for pre-flight tests, modeling and simulation, and other specialty ground testing will continue to mitigate flight test risk.

MDA has also taken the initiative to create a combined test force approach for GMD that has proven successful in other military system developments. The integration of contractor and government developmental and independent operational testers has created a team that is working together to plan, conduct, and assess the results of major GMD testing. This approach permits early operational tester involvement while still focusing on the examination of technical and design issues as the system matures through research and development. I will rely on this team as a primary information source in preparing my assessments, which will support Senior Executive Council and Defense Acquisition Board transition and deployment decisions.

In order to effectively and efficiently meet these new, as well as traditional, reporting responsibilities, I require extensive access into MDA's activity. This will be accomplished at three levels: first, through the Missile Defense Support Group that Secretary Aldridge has established; secondly, through regular communication with the MDA, and third, through my staff's direct involvement at the program and element level. In that regard, let me briefly discuss what has become referred to as "unfettered" access.

General Kadish and I both testified, as recently as two weeks ago, that DOT&E is being provided the necessary access to MDA programs. For example in the Patriot Advanced Capability 3 program, which is in its operational testing phase, we are involved on a daily basis. Due to the level of activity during the restructuring, we are not yet as involved in other programs.

Mr. Chairman, let me say that we are not yet where we want to be. However, the situation is improving, and General Kadish and I are committed to making it the best it can be.

I would be pleased to answer any questions you may have.

Mr. SHAYS. Thank you.

Dr. Stansberry.

Mr. STANSBERRY. Good morning, Mr. Chairman and members of the committee. Thank you for the opportunity to appear here today. I plan to address the new management structure and the more flexible oversight process adopted by the Department of Defense for its missile defense programs.

On January 2 of this year, Secretary Rumsfeld issued direction for the missile defense program. His stated objectives included the establishment of a single program to develop an integrated ballistic missile defense system under the authority of a single organization. He directed a capability based requirements process and streamlined oversight to facilitate the earliest possible deployment of missile defense capabilities. These changes are necessary due to the magnitude of the BMD program and the high priority placed on it by the President.

The Secretary will look to the DOD Senior Executive Council [SEC], for oversight and recommendations for decisionmaking in this area. The Senior Executive Council is chaired by Deputy Secretary Wolfowitz and includes the Service Secretaries and the Under Secretary of Defense for Acquisition, Technology and Logistics.

In response to the Secretary's direction of January 2nd, the Under Secretary for Acquisition, Technology and Logistics, Mr. Aldridge, issued implementation guidance. He directed the Director, Missile Defense Agency, General Kadish, to plan and execute a single missile defense program structured to integrate the work and enable capability trades across the different elements of the Ballistic Missile Defense System, and to facilitate decisive action in response to program events.

This new single program has the same reporting requirements to OSD and to Congress that all other DOD programs have. The Director, Missile Defense Agency will have the authority to manage the acquisition strategy, make program commitments to award contracts, to make affordability tradeoffs and to exercise milestone decision authority up to but not including milestone C, which is the beginning of the production and deployment phase.

When an individual element of the Ballistic Missile Defense System progresses to the point of demonstrating militarily useful capability, the Director of Missile Defense Agency will recommend that the Senior Executive Council consider it for transition to production and deployment. This transition will create an acquisition program in its own right and activate the management, oversight and reporting processes used for traditional defense acquisition programs, leading to a milestone C production decision by the defense acquisition executive advised by the Defense Acquisition Board. Then the designated military service will manage the program following standard acquisition processes and reporting.

To advise the Director of the Missile Defense Agency on management of his program and to aid the Senior Executive Council in decisionmaking on missile defense, a Missile Defense Support Group representing 13 selected staffs within the Department was formed. The Missile Defense Support Group consists of senior experienced individuals who provide useful insights and recommendations on policy, operations, acquisition and resource matters that affect the Ballistic Missile Defense System.

Over the past 4 months since the Missile Defense Support Group was established, it has met 10 times. This is a significant increase in the commitment of senior leaders of the Department compared to other DOD programs.

The Department is making these changes in management and oversight in response to the high priority for missile defense articulated by President Bush. We believe the changes will allow missile defenses to be developed and deployed in a much more efficient manner than otherwise. Congress has the same visibility in the missile defense program that it has with other DOD programs.

In this context, some of the classical metrics of progress will be affected by our approach to combine all research, development, testing and evaluation for missile defense into a single program that will be separate from programs for production and deployment. To ensure that Congress has a full understanding of the program, we are committed, committed to provide necessary details of how the program will be structured and managed.

The Missile Defense Agency has already conducted over 40 hours of briefings on the Ballistic Missile Defense System to congressional staff since the Missile Defense Agency was created, and will continue to provide Congress with detailed information to satisfy congressional oversight responsibilities.

Thank you, Mr. Chairman. I would be happy to answer any questions that you or the committee may have.

[The prepared statement of Mr. Stansberry follows:]

STATEMENT OF

DR. KENT G. STANSBERRY
DEPUTY DIRECTOR, MISSILE WARFARE FOR THE
UNDER SECRETARY OF DEFENSE
(ACQUISITION, TECHNOLOGY, AND LOGISTICS)

BEFORE THE
HOUSE COMMITTEE ON GOVERNMENT REFORM
SUBCOMMITTEE ON NATIONAL SECURITY, VETERANS
AFFAIRS, AND INTERNATIONAL RELATIONS

July 16, 2002

National Missile Defense: A New Organization, Evolutionary Technologies, and Unrestricted Testing
Dr. Kent G. Stansberry
Deputy Director Missile Warfare for the Under Secretary of Defense (Acquisition, Technology and Logistics)

Good morning, Mr. Chairman, Members of the Committee. Thank you for the opportunity to appear before you today to discuss the Missile Defense Program. My goal is to address the new management structure of the Missile Defense Program and discuss the new, more flexible oversight process which is being adopted by the Department for this program.

On January 2 of this year, Secretary Rumsfeld issued direction to the Department for the Missile Defense Program. His stated objectives included the establishment of a single program to develop an integrated ballistic missile defense system under the authority of a single organization, the Missile Defense Agency (MDA). He directed that a capability-based requirements process be adopted and that streamlined oversight be incorporated to facilitate the earliest possible deployment of missile defense capabilities to the Services.

Redesignation of the Ballistic Missile Defense Organization as the Missile Defense Agency provides greater authority to LtGen Kadish and his staff to manage the rigorous technical challenges associated with developing missile defenses. The additional authorities are necessary due to the magnitude of the program, and the high priority placed on this effort by the President. It is for these same reasons that the Secretary directed the use of a streamlined oversight process. The Secretary has indicated his intention to look to the DoD Senior Executive Council (SEC) for oversight and recommendations for decision-making in this area. The SEC is chaired by Deputy Secretary Wolfowitz and includes the Service Secretaries and the Under Secretary of Defense for Acquisition Technology and Logistics.

In response to the Secretary's direction of January 2, the Under Secretary for Acquisition, Technology and Logistics, Mr. Aldridge, issued implementation guidance. In this guidance, he directed the Director, MDA, to plan and execute a single Missile Defense Program, structured to integrate work and enable capability trades across different elements of the Ballistic Missile Defense System (BMDS) and to facilitate decisive action in response to program events. This single program has the same reporting requirements to OSD and Congress that all other programs have. The Director, MDA, has been given full authority to execute a capability-based acquisition approach that will produce missile defenses at the earliest feasible date. He will have the authority and responsibility to develop all associated technologies and conduct developmental testing. He will interface with the warfighter community to determine desired operational features and to develop strategies for introducing new capabilities to the fighting forces. He will have the authority to manage the acquisition strategy, make program commitments, award contracts, make affordability tradeoffs, and exercise milestone decision authority up to, but not including, Milestone C (the beginning of the production and deployment phase).

The unique management and oversight processes described above apply only to the development phase, when the configurations of missile defense systems are still being defined and the production and deployment considerations are unknown. When an individual element of the BMDS progresses to the point of demonstrating useful capability, Director, MDA, will recommend that the SEC consider it for transition to production and deployment. Transition to production will create an acquisition program in its own right and activate the management, oversight, and reporting processes used for traditional defense acquisition programs. We will establish the necessary product teams and processes needed to support a Milestone C production decision by the Defense Acquisition Board (DAB) to support this transition. Following the Milestone C decision, the designated Military Department will manage the program following standard acquisition processes and reporting.

To advise the Director, MDA, on management of the Ballistic Missile Defense program and to aid the SEC in executive decision-making on missile defense, a Missile Defense Support Group (MDSG) of designated senior experts from 13 selected staffs within the Department was formed of which I am the Executive Secretary. The Chairman of the MDSG is the Director of Strategic and Tactical Systems. He reports directly to the USD(AT&L) on all MDSG matters. The MDSG consists of senior and experienced individuals who provide useful insights and recommendations on policy, operations, acquisition, and resource matters that affect the Ballistic Missile Defense System (BMDS). Members of the MDSG are supported by a Working Group of individuals from their support staffs. They promote the success of the BMDS by providing informed constructive advice on the program and keeping their principals advised of program progress and developments. They perform independent reviews and studies of the Ballistic Missile Defense Program where needed. Over the past four months since the Missile Defense Support Group was established, it has met ten times. This is a significant increase of the commitment of senior leaders compared to DoD programs due to the magnitude of this program and the high priority placed on this effort by the President.

A significant task of the MDSG was to review the MDA plan to implement Secretary Rumsfeld's January 2, 2002 Missile Defense Program Direction, and the implementation guidance issued by the Under Secretary Aldridge. This plan addresses matters of great importance to the success of the BMDS, including the program structure and funding allocations, the management structure, acquisition strategy, program documentation, and interfaces with the Services. The MDSG reviewed numerous drafts of the implementation plan, and the MDA has incorporated the comments.

The Department is making these changes in response to the high priority for missile defense articulated by President Bush. We believe that integrating several programs into one, centralizing their management within a single defense agency with greater authority, responsibility, and flexibility, and providing a more streamlined oversight process will cause missile defenses to be developed and deployed in a much more efficient manner than would be possible under the former structure.

Congress has the same visibility into the Ballistic Missile Defense Program that it has with other programs. In this context, some of the classical metrics of progress are affected by our approach to combine all Research Development Testing and Evaluation (RDT&E) for missile defense into a single program. This is separate from any programs for production and deployment. To ensure that the Congress has a full understanding of the program, we are committed to provide necessary details of how the program will be structured and managed.

The MDA has already conducted over 40 hours of briefings on the Ballistic Missile Defense System to members of the Congressional Staff since MDA was created. Those briefings provide specifics on the planned development activities for each element of the system this year. MDA will continue to provide Congress with detailed information to satisfy its oversight responsibilities. The annual Selected Acquisition Report (SAR) to Congress was submitted on the BMDS with a separate SAR for the PAC-3 program, because procurement activity for PAC-3 has begun. Providing separate SARs will ensure full visibility into the transition of the program elements to the Military Departments after a deployment decision, just as it will be done for the Army's PAC-3.

Thank you, Mr. Chairman. I would be happy to answer any questions you and the Members of the Committee might have.

Mr. SHAYS. Thank you.

This is what we're going to do. We're going to do 5 minute questions to start; 5 minute questions aren't my favorite, because frankly you can't really get into great depth. I'll extend time if the witness is giving a longer answer. So I'm telling this to the Members, you don't need to interrupt the witness, we'll just extend time. So this way you get a full answer, and if we go 6 or 7, that's all right. Then we'll do a second round.

We are not going to leave this panel until we have answered the questions that need to be asked. So I don't want any Member to say, if we had another round we would have narrowed this down better. So we're going to go three rounds or four if we have to. Actually, the second round may be a 10-minute round.

Let me call on you, Mr. Gilman, to start.

Mr. GILMAN. Thank you, Mr. Chairman.

General Kadish, how confident are you that the missile defense system under development be able to distinguish between real warheads from decoys?

General KADISH. I'm becoming increasingly confident that that problem in the midcourse especially can be solved, and would be greatly complicated for our adversary by the additional layers of defense, especially the boost phase. So longer term, the evidence is building that we can handle this problem. But we're not there yet.

Mr. GILMAN. But you are working now. And General Kadish, what types of international cooperative programs were prohibited under the Anti-Ballistic Missile Treaty, and what are the technical benefits of working with our allies on missile defense?

General KADISH. Sir, although we had some very active programs at the shorter range missile problem with the Japanese and the Italians and the Germans for instance, and certainly the Israelis, the Treaty specifically prohibited us to share what we call blueprint data against longer range missile with our allies. So now that the Treaty is no longer in force, we are able to do that should we so desire.

Mr. GILMAN. So we're working with our allies now?

General KADISH. We are entering into discussions with our allies to see what might be the possibility in the cooperative area.

Mr. GILMAN. I note that GAO has made a recommendation that the agency use its opportunity to make its acquisition progress more knowledge based by establishing knowledge based decision points at key junctures. That would give the agency a better position to decide whether to invest in the next phase. Are you following that GAO recommendation?

General KADISH. For all intents and purposes, I believe we are. We have, for instance, structured the program such that on a yearly basis, at a minimum, the senior departmental leadership will look at our progress to date and then make decisions on the program, both in the RDT&E area and decisions to go forward based on that progress. So that in a larger sense, I believe, if I read the GAO's approach, meets the intent of what they're suggesting.

The whole program is structured so that we proceed based on our progress, not based on a schedule. And that's key to the event based approach that we're taking.

Mr. GILMAN. One further question, General. GAO said in their July 12th report they recommended the Director of the Missile Defense Agency not only establish decision points to separate technology development from system integration, system integration from system demonstration, and system demonstration from production, but also establish knowledge based criteria that would be used to determine whether additional investments should be made in the airborne laser program.

Are you following that recommendation to some extent?

General KADISH. Yes. I believe we are. We have basically three points that we are looking at for all our program elements. One is, the first point would be to enter into RDT&E. Second point is to transition from RDT&E into an early initial production. Then the third point is, as Dr. Stansberry pointed out, the full product transition to the services.

Each one of those points, we certainly will need criteria to proceed. And in the case of the ABL, I'm looking at specific criteria to use internal to MDA to support those types of decisions, not only at that decision point but prior to it.

So the baseline answer to the question would be, we're certainly proceeding in the direction that the GAO suggests. Maybe not exactly in the detail that they suggest at this point, because we're not quite there yet in terms of the schedules.

Mr. GILMAN. Thank you, General. We wish you success in all your endeavors.

Thank you, Mr. Chairman.

Mr. SHAYS. Thank you, Mr. Gilman.

Mr. Kucinich, you have the floor.

Mr. KUCINICH. Thank you very much, Mr. Chairman, members of the committee, General, welcome.

You showed us a very interesting video which appears to have been a success. I would just like some answers to the following questions so we could move this along. If a single answer of yes or no would suffice, that would be fine.

On those particular tests, did the target have a homing beacon or GPS technology?

General KADISH. The target had the range safety requirements to have a transponder on the target so we knew where it was going.

Mr. KUCINICH. So the answer is yes.

General KADISH. And we had a situation where the interceptor did not rely on those types of data for the intercept.

Mr. KUCINICH. Did the defense system then have advance information?

General KADISH. Absolutely. We have advance information on all our early RDT&E——

Mr. KUCINICH. So the system knew the speed of the incoming missile?

General KADISH. Absolutely. You need to have the truth data to compare it to.

Mr. KUCINICH. It knew the trajectory? Did it know the launch time?

General KADISH. Knew the launch time exactly.

Mr. KUCINICH. Did it know the launch location?

[Witness nods in the affirmative.]

Mr. SHAYS. Let me just say, the nodding of the head doesn't get us what we need for the transcriber. We just need a yes or a no. And you will be allowed to elaborate, just to clarify, and the gentleman will be given more time.

General KADISH. OK.

Mr. KUCINICH. Thank you, General. Thank you, Mr. Chairman. Did it know the trajectory?

General KADISH. These are early developmental tests, Congressman. We have a very stylized approach. They are very complex. We probably have anywhere from 15 to 20 specific sensor platforms looking at this. And we want to compare what these sensor platforms saw versus what actually happened.

Mr. KUCINICH. Thank you, General. Did it know the trajectory?

General KADISH. Yes.

Mr. KUCINICH. Did it know the launch time?

General KADISH. Yes.

Mr. KUCINICH. Did it know the launch location?

General KADISH. Yes.

Mr. KUCINICH. Did it have any advanced information about the decoy?

General KADISH. We have, within the mechanization of the system, what an intelligence activity would have, yes.

Mr. KUCINICH. There was a global positioning system transmitter in the first NMD intercept test. It provided the target location to the interceptor, correct?

General KADISH. It provided the target location to the people running the test.

Mr. KUCINICH. But it provided it to the interceptor booster to allow it to dispense the EKV in the correct place, did it not?

General KADISH. We have GPS, global positioning, to aid our test infrastructure. We don't use that information to do the integrative part of the test, however.

Mr. KUCINICH. Did or did not the test use anti-simulation or attempt to disguise the signature of the warhead in any way?

General KADISH. The decoys that we used are a class of decoys. They do not exactly replicate the warhead that was in the test. But that's for advanced data gathering for later tests.

Mr. KUCINICH. The test then was not one of discrimination, since it relied on the defense knowing in advance that the re-entry vehicle would be the object of the smallest infrared signal, isn't that correct?

General KADISH. The tests were designed to prove hit to kill, not to do counterdevelopment.

Mr. KUCINICH. In a real attack, the defense would not know in any detail what the warhead would look like, is that correct?

General KADISH. I would not say that.

Mr. KUCINICH. And the test, instead, wasn't this really a test of how sensitive the sensors are, and of the algorithms used by the NMD system and of the kill vehicle's ability to home in and hit a target?

General KADISH. That's correct.

Mr. KUCINICH. And while it's necessary to test algorithms, that's not the same as testing the discrimination ability of the kill vehicle, is it?

General KADISH. Well, it's an early look at what it can do. But it's not as robust as what we intend to do or what we need to do.

Mr. KUCINICH. But is the testing of algorithms, is that the same as testing the discrimination ability of the kill vehicle?

General KADISH. Yes.

Mr. KUCINICH. It is the same?

General KADISH. The algorithms determine, we might want to have a little tutorial on what the algorithm really is. But it is the mechanism with which the system determines what the target is.

Mr. KUCINICH. But it's not the same. When you test an algorithm, it's not the same as testing the discrimination ability of a kill vehicle.

General KADISH. Actually, I'd rather take that question for the record. It's a lot more complicated than you suggest.

The algorithms form the basis for the knowledge of the kill vehicle to do its job. So that's how you discriminate. You have the sensor data, the raw data from the sensor being worked on by algorithms that comes out with an answer.

Mr. KUCINICH. But when you're testing algorithms, you know, in one part of this first test, you were testing algorithms, you weren't really testing the discrimination ability of a kill vehicle.

General KADISH. Can I take that for the record? It's a lot more complicated than you suggest.

Mr. KUCINICH. OK, then, let me just conclude with this, Mr. Chairman, because my 5 minutes have expired. I think it's very helpful to, when you come before a congressional committee and you put these videos up, which show interceptors homing in on a target, I think it's very useful for our committee to qualify what we're seeing. We need you to do that when you're showing the video. Thank you.

Mr. SHAYS. I thank the gentleman.

General KADISH. If I could just comment.

Mr. SHAYS. Sure. Definitely, you have the floor.

General KADISH. There is no way, the intent of those videos was to in no way misrepresent our testing in any way, shape or form. It was only intended to show that we've answered the first question of three, as I stated in my testimony.

Mr. KUCINICH. If there is anybody in this room who heard the General qualify what we saw when we were seeing it, raise your hand, because I missed it, and I apologize if he said it.

Mr. SHAYS. Let me just say this to the gentleman, if I could, Mr. Kucinich. We intend to fully vett this whole issue. There are things that one may want to criticize General Kadish on, but I don't think he's attempting to suggest that it goes beyond what we saw.

Mr. KUCINICH. Would the gentleman yield?

Mr. SHAYS. I'd be happy to.

Mr. KUCINICH. That's fine, Mr. Chairman, but we have a room full of people here who were shown intercept tests. And it's important to evaluate them in the context in which we're seeing it. We can't do that unless we're given qualifying information about what we're seeing, that's all.

Mr. SHAYS. And let me just say to you that the only people that matter, frankly, aren't our guests, but the members of the committee. I think the General knew that we would fully question on this issue. I am not pressing him for time, but there are probably a lot of other things he could have said to qualify that video.

Mr. KUCINICH. Mr. Chairman, with the greatest respect, this whole discussion centers around the reliability of the tests. And if we're given information, and watching videos, which changes the context of this, it's important, I think, for us to be given information that would qualify what we're seeing. And we weren't given that. So I just wanted to make that a matter of record.

Mr. SHAYS. OK. And I'll use my prerogative to have the last word, just to say that we are fully going to allow you and the other Members to ask any question that has been shown. We are not going to let this panel leave before we think we have made sure that all sides and elements of the question are aired. That's just my point.

Mr. Putnam, you have the floor.

Mr. PUTNAM. Thank you, Mr. Chairman.

General Kadish, good morning. I was probably in high school physics more recently than anyone else in this panel, but help me out a little bit here. [Laughter.]

The target vehicles in the videos that you showed earlier in the hearing, how fast were they traveling?

General KADISH. The closing velocity was about 4.5 miles per second. I don't know the exact speed of each one of the objects, but their closing velocity was of that magnitude.

Mr. PUTNAM. And how far apart in distance was the target launch and the interceptor launch?

General KADISH. About 4,500 miles.

Mr. PUTNAM. So we intercepted a target traveling 4.5 miles per second, that was launched 4,500 miles away successful four of six times?

General KADISH. That's correct.

Mr. PUTNAM. Does that happen routinely?

General KADISH. Not to my knowledge.

Mr. PUTNAM. Is this a common occurrence in military research and design?

General KADISH. No, sir. We had some high profile failures in our attempts to do that. You might recall the first test was successful, but the next two failed, and we had some work to do to make it happen. The idea that we could actually do hit to kill in the atmosphere, in the trans-atmosphere, in outer space, more than hitting a bullet where the bullet is I think pretty proven, and not a regular course in the history of these types of weapons.

Mr. PUTNAM. Thank you. Isn't it also true that in cases where you have a failed test, there is success in the sense that the systems continue to test and monitor and provide data on what failed and which portion, boost, midcourse and terminal, and be able to correct those mistakes as a result of data that comes from the failure?

General KADISH. That's absolutely true and critical to our whole program. We have spent a lot of money trying to know what we call truth is, where the warhead actually was versus where it was supposed to be and those types of things. So that's a very important part of our test. And, however, it introduces artificialities that an operational test would not necessarily require.

Mr. PUTNAM. The other Members have quite correctly pointed out other examples of weapons systems that were very expensive that turned out never to be successful, never to be deployed. Is it fair to draw a parallel between your program and the space program, where there were very expensive, high profile failures leading up to the ultimate success and the ultimate deployment of sending men safely and returning them back home, to space?

General KADISH. I think we certainly have looked at many national programs in the past to come up and looked at how we can improve our management structure. I think it is a fair comparison that the technology regime that we're working in is unprecedented, and equivalent to some of the things we did in the past like the Apollo program and Gemini and that type of thing.

As you suggest, all our efforts along those lines had their failures and their successes. But ultimately, when we persevered, we were successful in the outcome.

Mr. PUTNAM. In terms of the different approaches, we talked about space based, we talked about sea based, land based, which of those platforms, in your opinion, and based on the current research, offers the greatest success with the soonest practicable deployability? Which one could be ready first and be the first line of what would be a successive wave of protections of this layered defense against missile threat?

General KADISH. Congressman, it depends on what range of missile you're talking about. For instance, against short range missiles, Patriot–3 is being produced and fielded in very limited quantities today. So that's the leading edge of the short range missiles.

When you go into the intermediate range, and medium range missiles, then the standard missile 3 Aegis, because we're two for two, and moving that program as rapidly as possible, is a leading candidate with the THAAD program, adding a ground based element to this hopefully soon after that.

Against long range missiles, our ground based missile defense system is the leading candidate, because we have tested it six times at this point.

So again, it depends on the range of the missile. It is important to understand that there is no one solution to the problem of missile defense, given the physics problem that we face. So a layered defense against all range of missiles is going to include an awful lot of those elements, in my view, to get the job done adequately.

Mr. PUTNAM. Thank you, and I look forward to the next round.

Mr. SHAYS. At this time the Chair would recognize Mr. Tierney.

Mr. TIERNEY. Thank you, Mr. Chairman.

General, let me reiterate some of the concerns that we have here. The previous system that we had allowed Congress some specific benchmarks on which to judge how the program was going. Under this new proposal, it doesn't appear that there is any way that Congress is going to be able to determine whether the program is over budget, because there are no cost projections. There is no way to determine an appropriate level of funding, because unlike other major weapons development programs, there is no final goal to which the program is striving. There is also no way for Congress to gauge the appropriate pace of the program. It can't determine whether progress is slow, because there are no dates by which requirements must be met.

In fact, it looks like there are no requirements at all. I just want to quote to you what the Congressional Research Service described in the lack of oversight. It said, a major consequence of the administration's proposed evolutionary acquisition strategy is that the missile defense program would not feature the well defined phases and milestones of the traditional Department of Defense acquisition system. Another consequence already reflected in the Department of Defense testimony is that the Missile Defense Agency cannot provide Congress with a description of its final missile defense architecture, the capabilities of any near or longer term system, the specific dates by which most elements of the emerging architecture are to be tested and deployed, an estimate of the eventual total cost of the missile defense program, or estimates of the amounts of funding that the program will require in individual years beyond 2002.

Now, just on the testing aspect of that, I know we had Phil Coyle, Mr. Christie's predecessor. He issued a report that we've already talked about, we had some difficulty getting it from the Department. But he described severe deficiencies in the testing program. He said the effectiveness was not yet proven, even in the most elementary sense. It was too immature to even assess its effectiveness in terms of the program. The program failed to test basic elements of the system, such as countermeasures and multiple engagements, which we expect to be the norm. And that the system would not be able to defend against accidental, unauthorized launches.

In all, he made about 50 specific recommendations. Can you tell us what the status of this program is in terms of its addressing those 50 very specific recommendations that Mr. Coyle made?

General KADISH. Mr. Tierney, the assessment that he made was against a program definition that no longer exists. He certainly had the characteristics of it and the legacy of it. But the basic architecture does not exist.

In fact, however, I believe most of the recommendations, and I have to get you that for the record, that he made are embedded in our concept of the test bed. We have put an awful lot of budgetary resources in place to get to the testing of the ground based system and then other layers as well, as we add them into a more rigorous and comprehensive set of tests.

And what we have done with the test bed is to address some of the main weaknesses that we all saw, as well as Mr. Coyle, in the way we were testing such a complex and unprecedented system. I believe that our current plans in fact address those weaknesses.

Mr. TIERNEY. You can understand how a skeptic may think that you just defined yourself out of a problem, that you have redefined your situation so as not to have to address those problems. So I'm going to ask if you would in fact provide us with the information of just how specifically those 50 issues are embedded in your new test program.

General KADISH. I'd be happy to, even though some of them might be totally applicable to the direction we're going now.

Mr. TIERNEY. I might as well stop here and come back on the other round.

Mr. SHAYS. OK, we'll do a 10 round match on that.

I would say, the gentlemen asked a lot of questions, I just want to make sure that when we put in on the table that we really go through them all. So I want to make sure you have the opportunity to do that.

I'm a little delinquent in not recognizing that Mr. Platts has joined us and Mr. Lynch. We thank you both for being here.

At this time, Mr. Schrock, you have the floor.

Mr. SCHROCK. I have no questions at this time, Mr. Chairman.

Mr. SHAYS. OK, the gentleman passes. So I go to you, Mr. Platts, if you'd like.

Mr. PLATTS. No, thank you, Mr. Chairman.

Mr. SHAYS. Mr. Allen, we'll go to you.

Mr. ALLEN. Thank you very much, Mr. Chairman.

I want to begin with Mr. Christie. We've heard a lot of testimony in the past about a test bed, the Fort Greeley test bed. In your testimony, you said, "When fully implemented, the test bed will mitigate many of the existing test limitations and restrictions that prevented operationally realistic flight testing identified during planning for the former national missile defense program."

My question is about what test bed we're talking about. General Kadish has testified that no flight tests are planned from the Fort Greeley site by the fall of 2004 in large part because booster segments would fall on populated areas. So help me with how you're using the word, and your implication that the test bed, Fort Greeley or something else, promotes operationally realistic testing of the ground based midcourse system.

Mr. CHRISTIE. I think the test bed involves a lot more than Fort Greeley. Now, it does involve the BMCQ setup there that will be used during testing.

Mr. ALLEN. I'm sorry, could you repeat that? It involves what?

Mr. CHRISTIE. It will involve the facility that will house the battle management command and control system that will be exercised during testing. What the test bed does with the other elements, in the longer term, is provide an opportunity to look at multiple launches, different geometries, both of which I think were criticisms that were rightly levied earlier by Mr. Coyle, as a matter of fact. In fact, the test bed will provide an opportunity to do quite a bit of the testing that were raised as issues.

Mr. ALLEN. Geographically, what are you referring to when you say the test bed?

Mr. CHRISTIE. Geographically it almost involves the entire Pacific Ocean. It will involve improvements at Vandenberg, Kodiak, perhaps the Shimia radar and other capabilities at Kwajalein, I think, that are not here now.

Mr. ALLEN. Second, in your testimony you mentioned the Senate provision that would further require TO&E to provide operational assessments for several missile defense programs. You state that you will satisfy the intent of that language under current reporting requirements. If you intend to comply with the spirit of the provision, what is the difference in complying with the letter of the provision that's in the Senate legislation, if that gets enacted? And would you object to the Senate legislation in the conference committee?

Mr. CHRISTIE. I do not object to the Senate legislation. All I'm saying is that I think the purpose of the Senate legislation will be served by my report next February and subsequent Februaries. I will include in those reports operational assessments of the system as tested up to that point.

Mr. ALLEN. General Kadish, I am not going to have as much time as I had hoped, but let me just ask you quickly, when you testified before the Armed Services Committee on June 27th, I asked a series of questions about the technological capability of the Fort Greeley, at least I thought I was asking about the Fort Greeley test bed, I know I was, by fall 2004. To the question, will MDA conduct a test by the fall of 2004 where the trajectory of the target is unknown, you responded that you had plans to, but didn't know whether the date was before or after September 2004. Do you know the answer to that question now?

General KADISH. After I got back, I checked. Right now we're planning it after that timeframe.

Mr. ALLEN. Likewise, are tests with unannounced target speed, launch time and countermeasure sweeps planned before 2004?

General KADISH. I checked, and that's subsequent to 2004.

Mr. ALLEN. Are tests involving a decoy mimicking the warhead, a tumbling RV and a radar jammer planned before fall 2004?

General KADISH. Some are and some aren't. I prefer to answer that more distinctly for the record, because some of that will be sensitive information.

Mr. ALLEN. That will be fine.

In your opinion, why was September 2004 selected for the date for emergency capability at the Fort Greeley test bed?

General KADISH. Implicit in the question that the emergency capability is the primary focus of the test bed, and the way we look at it is that the primary focus of the test bed is to do testing. What we call hardware in the loop, all hooked up, pan-Pacific test range, and that's the earliest we can do the test bed.

Emergency capability is something to be looked at over the next few years and subsequent to September 2004 to see whether or not that makes sense.

Mr. ALLEN. Mr. Chairman, if I could, when you use the word test bed, are you referring to Fort Greeley or something else?

General KADISH. Fort Greeley is a part, a major part, but only a part of the entire test bed. It's unfortunate that it's gotten the moniker of the test bed itself. Fort Greeley is the northern edge of the test bed. It extends to the Kwajalein, it extends to Hawaii, it extends to Vandenberg Air Force Base. So it would be helpful for us maybe to talk about the test bed in general, not just the Fort Greeley portion of it.

Mr. ALLEN. Thank you very much.

Mr. SHAYS. I'd like to ask some questions at this time. Then we'll go to our colleague from Massachusetts.

I am concerned that we walked through Mr. Tierney's statement, because he has a number of questions. He talks about benchmarks, and I appreciate him giving me his statement. He talks about benchmarks, funding, programs. Also just in terms of the pace of the program. Then what is not realistic is the question raised by Mr. Kucinich about what we're doing right now. I'm going to basically say that we have so much on the agenda, I don't know what we will be able to cover successfully here. But we haven't even gotten into what is the threat and are we responding to the threat. I am frankly going to put that off to another day.

But let me ask you in general terms to comment about benchmarks. What did we eliminate, in your view, and what replaces those benchmarks? And are benchmarks important, in your judgment? It's a more general comment.

General KADISH. Typically, in a program, the National Missile Defense Program, and I have to go back and check this, when we were specifying that, had some specific milestones it was supposed to meet in order to get to a 2005 timeframe. They were schedule driven and they were part of the process.

What we're trying to do now, however, is to look at what are the benchmarks or knowledge that we have, if you will, about this particular system as we progress over time. We will certainly have goals and objectives for a program. But this is not a program that has specific deployment goals yet, where we have specific quantities of interceptors that we're going to buy over a 20 year timeframe. We are still trying to decide whether or not we have enough confidence in the system to take it to the next step.

So in the next year to 2 years, and you will see this in the other elements, from ABL to Patriot–3, we have specific targets and objectives we are trying to meet with the dollars that have been allocated by the Congress in our specific program elements. But they are not the traditional, long term milestones that you would see with a major defense procurement program, because we're not to that stage yet.

We have deemed so far, at least at the MDA level, that we are not ready to procure other than the Patriot–3 system. So when we get to that point, we intend to have the benchmarks that we traditionally tend to associate with the defense procurement programs and go forward.

Mr. SHAYS. I'm trying to recall, in the 16 years that I've been here, I was thinking just parenthetically that when our colleague was talking about, he was boasting that he was the last to have high school physics, I made the assumption that probably when you were getting your first star, he wasn't even born. [Laughter.]

I'm a little troubled by this, because my 10 years on the Budget Committee kind of get me thinking that really, what you almost seem to be suggesting is that we inherently assume or know that there is a threat and we want this system. Those are the two things we know, and we know we can't deploy, because we don't have a system yet that works.

And because of that, we are less inclined to have benchmarks. It would strike me that we would, even if it's just an issue of research before we go to deployment, we would have certain benchmarks to be able to understand the whole issue of cost.

General KADISH. I apologize if I left you with a mis-impression. Certainly over the near term we have specific benchmarks that we are measuring to. Maybe Dr. Stansberry could comment on that.

Mr. SHAYS. So these are short term?

General KADISH. These are shorter term benchmarks, primarily because we are in an unprecedented technology area.

Mr. SHAYS. Let me just tell you what that says to me. What that says to me is that we do not have the ability to have long term benchmarks, because we simply don't know when this program is going to be workable. We're trying to have a bullet hit a bullet, basically. So we know that's pretty significant.

So before I get into the short term, you're basically agreeing that, or would you respond to my interpretation of what you're saying, and that is that we can't have long term benchmarks because it would be difficult to even begin to have them, given that we don't know when a system is workable?

General KADISH. We certainly have long term ideas of where we ought to go in the program. What I'm trying to suggest is that they are not as hard and fast as traditional, major defense procurement programs would have us have them.

Let me give you an example. We had those long term benchmarks for something called the Navy Area Program for a number of years in our portfolio. And as I look back on it, I think, I can get you the exact dates, but it seemed like we had three major restructures of that program every 18 months. All of those long term benchmarks changed because of the technology progress in the program. We weren't able to stabilize the program. We finally gave up on it in the process.

If we had paid attention to the shorter term program objectives, I think we would have probably made more progress in that program than we really did.

Mr. SHAYS. Yes. I'm not going to disagree with the logic, but the implications are significant. And let me just at this time recognize Mr. Lynch.

Mr. LYNCH. Thank you, Mr. Chairman.

Thank you very much, General, for your attendance here today. I'm new to this committee, and somewhat new to this issue. I did have an opportunity some months ago, however, to meet with a gentleman by the name of Theodore Postel at MIT, someone who is an expert in missile technology and I would describe him as a pretty strong critic of some of the proposals that you are putting forward.

In fairness, I just wanted to present some of his objections to you, just to get a sense of your response. After all, we're just trying to gather some information here. Professor Postel had much, much in line with what Mr. Kucinich was talking to earlier, had expressed the real difficulty in the technology that we were using, that at the very bottom of it it was unlikely that the current technology would ever be able to distinguish between an actual missile and a number of decoys that might be thrown up at the same time in a missile attack.

It was Mr. Postel's position that a very expensive missile defense system could be compromised by a technologically simple countermeasure. And if we're going to be putting all this money into such a system, we would want some assurances on behalf of the American people that this is going to in the end provide some significant measure of protection.

Where are we on this?

General KADISH. I guess the way I would respond is that countermeasures are always a problem for military systems, it doesn't matter whether it's missile defense or tanks or airplanes. In the case of missile defense, there are particular problems with countermeasures. But it depends on what phase you're trying to intercept the target missile in, whether it's boost phase, midcourse or terminal.

Now, certainly, each one of those phases has its own set of countermeasures and difficulties for the offense, but also there are difficulties for the adversary. The way I would respond is that our basic approach to countermeasures is the layered defense system that we're trying to build. Although we will probably end up building it incrementally, if we do so at all, what countermeasures work in boost do not work in the midcourse. Those that work in midcourse don't work in the boost phase. So the idea of layered defenses is its own counter countermeasure.

Certainly, within each one of those phases, we want to get as good as we can. There are techniques that we will use to make each one of those phases more countermeasure resistant. But the idea that we put forth to handle this problem is that we want a layered defense system that takes multiple shots at our adversary in each of the phases that the missile has to pass through. And that, we believe, will be the most effective counter countermeasure.

Mr. LYNCH. OK. That sounds like a very, very expensive system, and I'm not so sure—well, let me just go on to the second point that he had. And that was, I responded to him by saying that there had been some level of success in the testing, and that we had some results that indicated there was the ability here to at least some of these missiles down in the testing sequences.

His response was that the testing protocol that we're using was, well, this is also from some literature, from people who agree with him, that it was almost akin to skeet shooting, where in our testing, our defense systems were being told, here comes a missile. And so it's like, in skeet shooting, someone yelling pull, and then the target goes out and when we shoot it down. He said the basic premise that we're going to be instructed that there is a missile on the way is inherently faulty, and that it undermines the very results that we're being given.

Any response to that?

General KADISH. There is no doubt that the early developmental testing we have done, in the case of the ground based system that he's alluding to, we've only done six tests. We call it a walk before we run approach. You've got to understand how to pull the trigger on that shotgun before you know how to hit the piece you're looking for in skeet shooting.

So the early tests have been designed to be just that, early demonstration of the technology. There will in time, however, be more and more complications added to the test to get at this countermeasure problem in the midcourse. We certainly have not answered the question, how effective is this midcourse system, ground based, against the variety of decoys that it might go up against. We have not answered that question, and I don't think we will answer it any time in the short run. It's a long term problem that we're dealing with with that type of countermeasure situation.

That, however, doesn't mean that the system is ineffective. We have early indications from our study of this problem that we are gaining more and more competence, we will be able to solve the countermeasure problem in the midcourse to our satisfaction, enough to be competent in it. Then with the layering of the defense to complicate the adversary's problem, we would make that even more effective.

So that's the approach. And over time, only time will tell in our tests just how effective we will be against countermeasures.

Mr. LYNCH. General, I want to thank you for your honesty and your frankness here today. Thank you, Mr. Chairman.

Mr. SHAYS. I thank the gentleman.

This is what I'd like to see if we can do. We have an issue of benchmarking, we have one issue of how realistic are the tests now and where we have to go. We have issues dealing with the benchmarks, the funding of the program itself and the pace. I'm just wondering, and I'd be willing to allow unlimited time on this, if we could just talk about the realistic aspects of the test right now.

And if we could just find out what's realistic and what isn't realistic about the tests, and allow Members to just focus on that, I'm wondering if that would be agreeable. I want to participate in the questions. Mr. Putnam wants to be able to jump in, but doesn't need to necessarily reserve time.

Would that be agreeable to you? Why don't we just see how it goes.

So let me throw out the question, starting the first round, and allow you to interrupt, Mr. Kucinich, and take over, ask some questions as well. And you as well, Mr. Tierney, and you, Mr. Lynch.

Tell us, with what you showed us, tell us what is realistic, or what was done that makes it artificial to start but still an accomplishment. I'm going to consider a bullet hitting a bullet as significant. Knowing when the gun shot off, knowing what level, knowing all these things, it's a start. But eventually, in order to be realistic, it has to know more.

So let's answer the question really that my colleague Mr. Kucinich wants to know. What was realistic and what wasn't realistic about this test? Where do we have to go beyond this in order be to realistic? Or if you want to ask it a different way, I think you get the point. Why don't you jump in to answer that.

General KADISH. Well, Mr. Chairman, maybe we ought to define what realistic means, just for purposes of our conversation, if it's all right. Then we can talk about it later.

Mr. SHAYS. Fine.

General KADISH. Normally what people mean when they say realistic is to design a test so that it represents as close to what a war fighting situation would be, if we were under attack from North Korea or whatever location, how would the system respond to that attack that is unannounced, that has some decoys and those types of things in it.

We have not conducted that kind of test against any of our systems except the Patriot–3. And even in the operational tests of Patriot–3, there was some artificiality.

Mr. SHAYS. The Patriot–3 is the short term?

General KADISH. It's the short range missile.

So the films that I showed you were realistic only in the sense that we, from a technical point of view, set objectives to do hit to kill, something that's not been done before in those regimes, and to do it repeatedly enough to have confidence that we know how to do that.

Now, as we progress by building the test bed for the ground based system and our other activities, we're going to vary the geometries, we're going to make the tests harder from a developmental standpoint, to do full out what we call the envelope of its performance. Then we will at some point in this process make a determination that we should go to more operationally realistic testing, which is the thrust of your question.

I suggest that will probably, for the ground based system, be somewhere in the neighborhood of 2004, 2005, 2006 timeframe. And we will be doing it in concert with Mr. Christie's activity.

Today, we cannot do that type of testing to the extent we want to because of the limitations of our test infrastructure. When we build a test bed, we will have more opportunities to test in a realistic way than what we have today. That requires investment and it requires 2 or 3 years of time to build it.

Mr. TIERNEY. Mr. Chairman.

Mr. SHAYS. Before the gentleman jumps in, let me just ask, so I'm clear, are we talking three different types, we're talking the short range, mid range and long range? And with the Patriot–3, we're just talking short range, right. But you are already on record as responding to Mr. Kucinich and saying we obviously knew it was going to take off, we knew the level it was going to be, etc.

Excuse me, just jump in, General.

General KADISH. Those are the types of things we want to know in early developmental testing, so that if we have a failure or even for that matter if we have a success, we know exactly what happened in the timeframe. We try to reduce those variables in early developmental testing, so that when we go to realistic testing, we can be confident it will work as intended.

Mr. SHAYS. Let me just tell you what would be helpful, though. It would be helpful for you to be able, here or later on, to list all the things that are not "realistic" about it. And it's probably a fairly long list. It would seem to me that if you're saying those are the variables that you would just begin to tell us the things you will begin to add to make it more realistic. That would be helpful for us to visualize.

General KADISH. We will do our best.

Mr. SHAYS. Mr. Christie, are you capable of responding to this part of it?

Mr. CHRISTIE. Certainly. We will provide that for you.

Mr. SHAYS. Thank you. Do you want to jump in?

Mr. TIERNEY. I do, Mr. Chairman, if it's all right with you and the ranking member. I'm going to take this slightly in a different direction, but not much. We used to have an operational requirements document. In fact, that's what Mr. Coyle used, it was very helpful to him to make an assessment of where the program was, and whether or not it was achieving its stated goals at a given time.

My understanding of the ORD is that it set forth specific objectives, specific military requirements and specific time lines. Now Secretary Rumsfeld has done away with the ORD. So how are you going to communicate to Congress what this program's specific objectives are, its specific military requirements and its specific timelines?

Mr. SHAYS. Could I just ask, just to know, I'm happy to spend as much time as we need to on that issue, but are we content with leaving the whole issue of realistic?

Mr. TIERNEY. Well, there's no chance it's even close to realistic. I don't know how much you want to explore that.

Mr. SHAYS. I just want to make sure we deal with that issue. You can have 15 or 20 minutes to pursue your question.

Mr. TIERNEY. I don't think you can break it up that way, but I'm fine to go back and give it a shot.

Mr. SHAYS. Let's just see if we can know what's realistic or not. Did you want to do that?

Mr. KUCINICH. First of all, I'm pleased that Mr. Tierney, who has done so much work on this, continues, from my point of view. But I am going to go back to the testing and what's realistic, whenever that's appropriate.

Mr. SHAYS. Yes, go ahead.

Mr. TIERNEY. I don't think you can break it down that way and make any sense, but I'm happy to yield.

Mr. SHAYS. Let's just experiment and see, even though we don't have a long range plan here. This is about what's realistic.

Mr. KUCINICH. General, do you believe that a judgment that the planned missile defense system can work against realistic countermeasures must be based on a sound analysis of the performance of the planned system against feasible countermeasures designed to defeat it?

General KADISH. As you state it, I don't have any problem with that at all. That's what we're trying to do.

Mr. KUCINICH. Should such an analysis indicate that the NMD system may be able to deal with such countermeasures, do you believe that there should be a rigorous testing program that incorporates realistic countermeasures created to assess the operational effectiveness of the planned system?

General KADISH. I do. We have a small problem in approaching that, because we have to build those countermeasures, and that's not a trivial task.

Mr. KUCINICH. Do you believe the United States should demonstrate that the system could overcome such countermeasures before a deployment decision is made?

General KADISH. I think that there are many ways you could decide to deploy, but I would certainly want to have some personal confidence that we can handle countermeasures that we expect to see.

Mr. KUCINICH. So is that a yes, then?

General KADISH. It is as I stated.

Mr. KUCINICH. I will ask it again. And it's true, I got the answer, and it's not very clear.

I want to ask, because it may be difficult or impossible to obtain direct information about the countermeasure programs of other States, do you think that the United States should rely on red team programs that develop countermeasures using technology available to emerging missile States to assess the countermeasure capabilities of potential attackers?

General KADISH. I not only believe that, we have set that structure up. It has actually been over a year, I think, in operation, using that technique.

Mr. KUCINICH. Isn't it true, General, that the red team programs that currently exist are under the financial control and authority of the Ballistic Missile Defense Organization?

General KADISH. That's true.

Mr. KUCINICH. And isn't that, General, a conflict of interest, inherently?

General KADISH. I don't believe so.

Mr. KUCINICH. I don't have much military experience, but I know from brothers who have served that when they take part in maneuvers and there's a red team, that red team doesn't have information about what the other teams are doing.

General KADISH. We have set up a structure that General Larry Welsh chairs our red, white, blue team structure that you're suggesting there. Although it's true we have chartered that activity, the access to information by the two gentlemen sitting on either side of me is certainly going to be there. I can also attest to you, you might want to call General Welsh. But he is no shrinking violet in his opinions, and in fact has been one of our more constructive critics over time.

So I am very confident that the independence you suggest is inherent in the program, even though we are responsible to make sure that our countermeasure program is——

Mr. KUCINICH. And I want it said for the record that the General has testified that the red team programs are under the financial control and authority of the Ballistic Missile Defense Organization. These are people that are supposed to be giving an independent evaluation of whether or not these countermeasures work.

Now, I want to go over a few points here about the testing program. Do you believe that the testing program must ensure that the baseline threat is realistically defined by having the systems threat assessment requirement, or the STAR document reviewed by an independent panel of qualified experts?

General KADISH. I don't think that's going to be good enough, Congressman. I think that we certainly need to pay attention to the intelligence based description of the threats. Because that's an important element in our construct.

But inherent in the approach we're taking, capabilities based is that I don't want to trust that document. I want to go beyond——

Mr. KUCINICH. You don't want to trust what document?

General KADISH. The STAR document. It's certainly an important piece. But from a physics standpoint, there are many things that we need to do to make our systems responsive, that we could make inherent in our systems that don't pay attention to an exquisitely defined threat. Because if we go that route, we will always be chasing the threat.

What our intention is is to put a capability in the system, so that the threat that the STAR and other documents from the intelligence community produce falls within that space so we can handle it.

Mr. KUCINICH. Thank you, General.

Mr. Chairman, I raised that issue because there is available evidence that strongly suggests that the Pentagon has greatly underestimated the ability and motivation of emerging missile states to deploy effective countermeasures. There are strong indications that the Pentagon's STAR document, an operational requirements document, which describe the type of threat the NMD system must defend against, underestimates the effectiveness of countermeasures that an emerging missile state could deploy, and inaccurately describes the actual threat. If the threat assessment and requirements documents do not accurately reflect the real world threat, then an NMD system designed and built to meet these less demanding requirements will fail in the real world.

So that's why, General, I asked the question, if the testing program would conduct tests to ensure that the baseline is realistically defined by having the STAR document reviewed by an independent panel of qualified experts. Mr. Chairman, it's a central question that relates to the real world.

General KADISH. Congressman, I agree with you. I don't think the STAR is adequate, either.

Mr. KUCINICH. Then I'm going to ask you, General, whether the testing program must conduct tests against the most effective countermeasures that an emerging missile state could reasonably be expected to build?

General KADISH. I agree with that, too.

Mr. KUCINICH. And I'm going to ask if the testing program must conduct enough tests against countermeasures to determine the effectiveness of the system with a high degree of confidence?

General KADISH. Not yet, Congressman, but we intend, when we move into test bed, to do just that.

Mr. KUCINICH. And finally, for this round, I want to ask if the testing program must provide for objective assessment of the design and results of the testing program by an independent standing review committee?

General KADISH. We have a committee in front of you today, I guess, plus a lot larger part of the DOD. I'm not sure what you mean by independent, but I'm sure and confident that it will be reviewed by the decisionmakers with many different points of view.

Mr. SHAYS. I thank the gentleman.

What would be helpful for the committee to have would be all the variables that ultimately add up to a realistic test, and for me, I don't have any trouble understanding that right now it is not realistic, because we're in the infant stages of this. But eventually, we will have to go through all of these different variables. I would be eager to know if the administration would be advocating going to production before we get to all those variables, whether if we get to three-quarters of the variables are they going to start to say, we need to go into production.

So it's not something I'm going to ask for now, because I don't think we're going to answer all the questions. But what's clear to me, what I've always assumed was the tests were never realistic because we're not at that point yet. And I'd also like to know, even though we know we don't have these benchmarks, you said 2004, 2006, I'd like to have it a little bit more pinned down.

I'm going to also say that we've had other administrations, Republican and Democrat, who have come before this committee with benchmarks that all of us know are a joke. That's about as unacceptable, even more unacceptable than no benchmark. At least it's an honest response that we aren't there yet. But it does say that we have then no ability to even begin to know its cost.

What I'm going to do is, I'm going to give the gentleman, just for a sense of time, I'm happy to give him 10 minutes and I'm happy to roll over 10 minutes. But I just want to have a sense of time, and give you the floor.

Mr. TIERNEY. Other than costs, setting that aside, I think it's important to have these benchmarks where you make a determination of whether it's worth proceeding forward at some point in time, or when you change direction or when you give up.

Does the chairman intend, when you asked for a list of realistic factors, to include also whether or not the tests will be using the actual equipment that will be anticipated for use in the final product, versus whether or not the booster will be the final booster versus the prototype, whether or not you'll be using the x-band radar versus the Cobra, just updated, or is all that part of that, or are you only talking about whether and other variables?

Mr. SHAYS. Let me just say that there are some things that won't be totally realistic, because we won't have a fixed—we won't necessarily have a facility in Alaska set up, a facility in Great Britain set up, we won't necessarily have a final product. But to answer your question, I want to know all the variables that add up to what the Department thinks will be ultimately a realistic test. That's what I'm asking for.

Mr. TIERNEY. Just to finish up this realism issue here, right now, the interceptor that's being planned basically determines whether it's a weapon that is striking or a decoy, using the infrared system.

General KADISH. We intend to use not only that infrared onboard system, but the basic operation is that, the early stages of this is that the kill vehicle has a large burden in that, yes.

Mr. TIERNEY. So it's like looking at it?

General KADISH. Right. But there are other elements in the radar area that we want to mechanize as soon as we can.

Mr. TIERNEY. So it will more distance—we're a decade away from the x-band, right, at least?

General KADISH. I wouldn't share that opinion, no.

Mr. TIERNEY. Well, tell me what your opinion is, because I want to hold you to it some day. When do you think we'll be ready with the x-band?

General KADISH. It depends on what type of x-band you're talking about, Congressman.

Mr. TIERNEY. One that works. One that is anticipated to work.

General KADISH. I assume it will work. We're trying right now to figure out, without the ABM Treaty to put on restrictions, just what size x-band makes sense and where we ought to put it in our architectures. Because of the treaty limitations now going away, there are some possibilities opening up where we can use less powerful x-bands than what we were talking about 2 years ago, with the big radar at Shimia, for instance.

So I don't want to answer that question yet. But the indications I have are that it will be a lot sooner than you suggest.

Mr. TIERNEY. It may be, if those circumstances come to bear that you can use the smaller ones.

General KADISH. That's correct.

Mr. TIERNEY. You still have to negotiate use of land, wherever you do decide to locate, whatever States might be involved, right?

General KADISH. Well, sea based radars are not out of the question.

Mr. TIERNEY. Not out of the question, but land based ones aren't either, and it they're land based you've got do some serious negotiation.

General KADISH. Well, if we can do them on sea, we can do them on land and then it depends, as you suggest.

What we need to do now and what we're looking at very carefully is deciding what to put in our test bed in the Pacific to prove out these concepts. Because I don't want to only rely on our analysis, based on the testing we've done to date on this.

Mr. TIERNEY. One of the issues that comes up when you're talking about using infrared to identify or distinguish between the missile coming in is of course, it's looking at it, in a sense. And when you're in that space area, as you're talking about, there's no air drag. There being no air drag, it's very difficult to distinguish between slow and fast, as the weight isn't holding it up.

So there has been, to my knowledge, no test and no scientific theory even that tells us how we're going to be able to do that, if someone decides to surround a weapon with an envelope of some sort that makes it look like a decoy. I'm not even aware of any scientific theory, maybe you can enlighten me on that, but I don't know of any scientific evidence or analysis yet that shows we're able to deal with that situation.

General KADISH. Mr. Tierney, I would offer to you, I would be more than happy to come over and discuss in a classified forum why we think we can do that.

Mr. TIERNEY. OK. Is this something more classified than the last one you tried to show me about?

General KADISH. I don't recall trying to show you that.

Mr. TIERNEY. We had a classified meeting that I was at with the Armed Services Committee a while ago and we went through this adventure.

General KADISH. You're asking a very specific question about the kill vehicle, YR. We didn't discuss it in that level of detail at that session.

Mr. TIERNEY. Well, maybe we should meet then. I'd be happy to sit down and talk with you about that. Because you would agree with me that's a concern?

General KADISH. That's been a concern all along. That's why we're testing.

Mr. TIERNEY. I guess on this idea of testing, one of the reasons for this is, back in 1997 and 1998, the agency; I don't even think you were here yet at that time, when did you start?

General KADISH. I think it was 1999.

Mr. TIERNEY. In 1997 and 1998, the agency ran some tests and basically tried to tell us all what a great success it was. Then we all looked back, we looked at it and found out that it wasn't such a great success after all. They claimed success in distinguishing and they said it was beyond all expectations. Well, apparently it was, because the GAO found out that the sensors had failed, and that the claims of success could not possibly have been true.

So that's what I want to make sure of, that we don't get into that situation. I'm really concerned that we have some ability, that the GAO or a critic like Mr. Postel, or the Union of Concerned Scientists or other people be able to evaluate what's going on, other than the agency itself. You're not going to be there forever, so it's not bringing your integrity into question. But in 1997, 1998, we had that kind of scenarios happen. After that, we had a situation where 10 times brighter decoy than a bomb was used and it struck and everybody claimed success. Only later did they find out that it was preprogrammed to strike the brighter object, and distinguish it on that sense.

So I want to make sure that we have in place the kind of safegauards were people can look at it and see what's going on. Now, Mr. Coyle tells us that we're not anywhere near the development of decoy technology and situations, a matter where we should be classifying this stuff, and that we're a long distance away from that, we're still in the somewhat rudimentary stages here, that essentially we're just holding it from the public. That disturbs me.

When we get to a point, someplace down the road, there may be a time to classify stuff and keep it out of everybody's mind. But in the meantime, I think it's really important, if you're talking about spending $238 billion, eventually, which is numbers that we'll go over in a while, then I think the public has a right not just to have Congress look at these things, because sometimes Congress doesn't jump on board with anything military, but to have an assessment done by people. We're not all rocket scientists in the public, but some people are. They ought to have a right to look at it and make a determination and raise some constructive criticism and move forward on that.

So maybe you can share with me what's going on now that you think you have to classify that kind of information at this point of early development. And contrary to Mr. Coyle, who I would think is a fairly knowledgeable and sophisticated individual's opinion.

General KADISH. With all due respect to Mr. Coyle, I do not share his opinion on that issue. We are at a point now, as I tried to indicate with my description of the ops tempo testing, that is not outside the realm of possibility we might be proposing specific design construction of a system over the next few years, based on our testing progress. The testing that we start doing now with countermeasures in my view should start to be limited in terms of its access to the general public view, for a simple reason. That is that if we specify the exact type of countermeasure, its dimensions, down to a hundredth of an inch that we did on previous tests, the types of material it's made out of, how it's deployed, where it is in the constellation, then I think we would be derelict in our duty to prevent that information from getting in the hands of people who could use it against us.

Mr. TIERNEY. Yes, but you're not—come on, General, be serious, you're not even at the stage where you're moving outside of broad categories on this stuff. You're talking about rudimentary, you haven't even tested strike objects coming in, things of that nature.

General KADISH. We intend to make our tests more complicated. And as we do, I see no value in allowing our adversaries that will wish us harm to know that type of data with great specificity or confidence.

Mr. TIERNEY. Let's assume that our adversaries aren't totally dense, that they're at least going to assume some of the lower level types that you'll be testing on at these early stages. I mean, they've written about them everywhere, I've seen reports, you've seen reports, the world has seen reports on it. So I know and they know that you're going to do testing at that level. There is no reason to keep classified the results of your tests as to how you're doing on that, other than to hide failures.

And I think that's what disturbs me. You will be testing immediately some very simple, so it's decoys and testing on that basis, before you get down the road to any of the things you're talking about, more complex and complicated, which at that point you may want to go classified with it.

But in the interim, General, be serious with us and be blunt. You're talking about some very rudimentary things here, where there's no need to keep it classified, nor is there a need to keep classified the results of how your tests against them do.

General KADISH. Congressman, I just differ with your opinion on that. I believe it's time that we start being serious about this as a war fighting technology——

Mr. TIERNEY. Well, let me object to that for a second. Nobody's less than serious about this, General. Let's be adult about this and not start that kind of terminology. We're all dead serious about this. The fact that we're dead serious about it means that none of us wants to find out later that somebody's been pulling the wool over our eyes like they did in 1997 and 1998 and shortly after that.

So while we're here in these rudimentary stages or whatever, we'd like to see, as the American public, whether or not your tests will work, not to hear somebody say, hey, we're dead serious about it, so even at that stage, you can't find out. We want to trust you. But trust would mean that when you get to a stage where it's reasonable to go classified, that's when you make the recommendation to go classified, not at some rudimentary stage where everybody in the world knows what kind of decoys you're going to be testing against, they've read it in some literature and it's all the way out there.

General KADISH. Congressman, I don't know what to say other than, we will give you and the Congress full access to the results of the tests. There is no question about that.

Mr. TIERNEY. But not the American people.

General KADISH. And as elected representatives of the American people, you can judge whether that's appropriate or not.

Mr. TIERNEY. So you won't be classifying them.

General KADISH. We intend to keep very sensitive data classified.

Mr. TIERNEY. But you're telling me what you now are going to tell us is very sensitive is even your testing of very elementary types of decoys.

General KADISH. I believe it's time now to do that, yes, Congressman.

Mr. TIERNEY. I don't know how much longer you want me to go before somebody else gets a shot.

Mr. SHAYS. We're going to get to the next panel soon. But I want you to proceed with questions. Mr. Putnam wants to ask some questions.

Mr. PUTNAM. If I could, Mr. Chairman, just to pursue Mr. Tierney's line on this classification issue, can you name any major weapons system that has open or unclassified access in its early stages of development? Do we have the early stages of our stealth capabilities unclassified? Do we have our unmanned aerial vehicles and unmanned water vehicles, is that technology in its embryonic stages unclassified? Is there any major weapons system where we have had the embryonic stages of development and experiments and testing unclassified?

General KADISH. I don't know if I can answer that categorically, but in my experience, maybe Mr. Christie can add to it, is that we generally don't reveal our weaknesses to the public in our specific weapons systems in terms of their vulnerabilities. That's reserved for internal decisionmaking and for the Congress to decide.

Mr. CHRISTIE. I would elaborate on that, but I think the issue is one of classifying capabilities or weaknesses against specific countermeasures. That's standard practice in the Department. We deal with classified information along that line all the time, whether it's an aircraft and it's the vulnerability of its radar to somebody's countermeasures or whatever. So without getting into the merits of whether this should or should not be done at this point in time, I don't think it's any different than what we've done in other weapons systems. We're talking about vulnerabilities——

Mr. PUTNAM. So systems vulnerabilities are routinely classified?

Mr. CHRISTIE. Yes?

Mr. PUTNAM. Systems vulnerabilities are routinely classified?

Mr. CHRISTIE. They have been in the past.

Mr. PUTNAM. It is not unusual or unique for missile defense vulnerabilities to be classified?

Mr. CHRISTIE. In my experience, no, that is not.

Mr. TIERNEY. Can I ask a question for a second, if you'll yield?

Mr. PUTNAM. I yield.

Mr. TIERNEY. Certainly it wasn't classified in the first several tests that we had that dealt with some forms of decoys. It only because of something that we wanted to classify when they were failures. In the earlier tests, decoys were used, the public was informed. If you look at the literature, there are volumes on it.

So I just want the member to know that you may have intended to be genuine in your response, but at least with respect to this program, they've not been classified, and are only becoming classified now at these early stages. And I would hope that the gentleman might inquire about the joint strike fighter, about the F–

22, about other weapons systems, all of which were tested publicly, which knowledge was not classified until much, much further down the road on that.

Mr. PUTNAM. Reclaiming my time, the joint strike fighter's vulnerabilities are unclassified, is that what you're saying?

Mr. TIERNEY. The tests. The testing for vulnerabilities at the very early stages of the decoy system. That will come at a later point when they're talking about more sophisticated types of decoys.

Mr. PUTNAM. But I believe that Mr. Christie's specific testimony was that vulnerabilities are routinely classified, is that correct?

Mr. CHRISTIE. That's right.

Mr. PUTNAM. I yield back, Mr. Chairman.

Mr. SHAYS. We're eventually going to get to the next panel. I do want to make sure, though, again, that we nail down the questions.

Mr. KUCINICH. We're nailing down our questions, Mr. Chairman, we're not nailing down the answers. And I would respectfully suggest that the line of questioning Mr. Tierney has been pursuing is probably the most important questioning that we've had in all these hearings. Because it really relates to what's legit and what's not. It relates to whether there's any fraud going on, frankly, or not. That ought to be a concern to the people.

Because if the planned missile defense system could be defeated by technically simple countermeasures, and it in effect has been, people ought to know that. Because then the whole system is in question.

So I think Mr. Tierney's question was right on the mark, and with all due respect, General, I'm going to give you another chance. The planned missile defense system, isn't it true that it could be defeated by technically simple countermeasures?

General KADISH. I don't agree with that characterization. I think we have a lot of questions to answer to in terms of what it can and cannot do with countermeasures. But the missile defense system we envision is a layered defense system. It has a boost component, it has a midcourse component, potentially it has a terminal component. Countermeasures that work in boost do not work in midcourse, those that work in midcourse do not work in boost. And an adversary's problem is greatly complicated by that type of a system.

Mr. KUCINICH. Isn't it true, General, that such countermeasures, simple countermeasures, would be available to any emerging missile state that deploys a long range ballistic missile?

General KADISH. Countermeasures are always available.

Mr. KUCINICH. And isn't it true there are numerous tactics that an attacker could use to counter the planned NMD system?

General KADISH. That's exactly why we don't want to reveal our vulnerabilities.

Mr. KUCINICH. And isn't it true that none of these countermeasures are new?

General KADISH. I would not characterize that at all. There are always new countermeasures. The F–22, the C–17, you name the system, has countermeasures today that they're facing, that they're vulnerable to.

Mr. KUCINICH. Isn't it true that most of the ideas that are countermeasures are as old as the ballistic missile itself?

General KADISH. Countermeasures are a fact of life in military systems.

Mr. KUCINICH. Isn't it true that all the countries that have deployed long range ballistic missiles, that is Britain, China, France, Russia and the United States, have developed, produced and in some cases, deployed countermeasures for their missiles?

General KADISH. I would assume so.

Mr. KUCINICH. General, you're in charge of this program. Are you assuming it or do you know it?

General KADISH. I know that there are people, including the United States, that have developed countermeasures for ballistic missile to make it more effective.

Mr. KUCINICH. Do you have any reason to believe that emerging missile states would behave differently than the countries that I've just articulated?

General KADISH. In the long run, if they had the ability, absolutely not.

Mr. KUCINICH. And do you believe that many highly effective countermeasures require a lower level of technology than that required to build a long range ballistic missile?

General KADISH. I'm not sure how to answer that question, because if you look at, I'm told by very reliable people, after looking at this in depth, that the countermeasure development experience by the United States was very difficult to make them effective. So there is a lot of opinions on that. And quite frankly, I have not made up my mind whether or not these are easier to do or hard to do. But I can tell you that trying to make them to test against, and looking at how we make countermeasures to test against, it's not a simple proposition.

Mr. KUCINICH. Can an attacker disguise a warhead to make it look like a decoy?

General KADISH. Sure.

Mr. KUCINICH. And when that happens, isn't it possible that the attacker could place a nuclear warhead in a lightweight balloon made of aluminized mylar and release it, along with a large number of similar but empty balloons?

General KADISH. Anything is possible.

Mr. KUCINICH. Well, if that's possible, then the defense would need to shoot at all the balloons to prevent the warhead from getting through, but the attacker could deploy so many balloons that the defense would run out of interceptors, isn't that possible?

General KADISH. It's possible, but if we had a boost phase layer, the chances are that we would have got that particular set of countermeasures before it was even deployed.

Mr. KUCINICH. And with respect to sub-munitions, isn't it, to deliver a weapon, whether it's biological or chemical, by a long range ballistic missile, an attacker could divide the agent, isn't that true, for each missile, among 100 or so small warheads or sub-munitions that could be released shortly after boost phase?

General KADISH. That's possible. It would be very interesting and difficult for them to do. But if we had a boost phase layer, like we're intending to have, that would be a countermeasure to that particular countermeasure.

Mr. KUCINICH. Wouldn't it be true that the sub-munitions would be too numerous for a limited defense such as the planned NMD system to even attempt to intercept all of them?

General KADISH. Our planned NMD system is a layered defense that includes boost, midcourse and potentially terminal systems. So I believe that if we had such a system, it would be effective.

Mr. KUCINICH. You testified earlier that this—you know what, Mr. Chairman, I'm going to pass right now.

Mr. SHAYS. OK. Let me just jump in and clear up what I need to clear up.

I accept the fact that testing will become more and more realistic as each test occurs, to a point where we'll know if the system works or not. And so I just intuitively accept that.

I am having a little bit of wrestling, and I actually accept the fact that it is difficult to do benchmarks when you really don't know when you're going to get a realistic system and when you're going to go to deployment. So I can understand why long term benchmarks become almost absurd.

But what I'm having a hard time understanding is, given that, how do we even begin to estimate costs? Whether the figure of $250 billion to half a trillion, I have no sense of how we can even begin to estimate costs. Do you think in fact, given that we don't have any benchmarks, given that we don't know yet if the system will work, is it possible to have a sense of cost, long term?

General KADISH. Mr. Chairman, that's a very important question that we struggle with an awful lot. I can tell you that as we are developing our budgets and our specific program approach, we will have as good as we are able to cost estimates for the R&D part of this. And we're developing those now and hopefully in the near term we'll be able to tell people what they are.

But they are indicated, for instance, in the level of budgets we have submitted to the Congress in the 5-year defense plan, with the amount of dollars we are allocating to this effort. Less than 2 percent of the budget, but about $8 billion a year. So you can get a sense from an R&D point of view, if we should execute that program at those funding levels, that somewhere in the neighborhood of $35 billion to $40 billion is where we're headed. Hopefully we can do it for a lot less.

When you go beyond that, for procurement, it becomes, and this is where it gets difficult, the question of how much force structure you want to buy, how many missiles, how many radars and how many ships, how many soldiers are required to do those types of things. Those decisions are very difficult to estimate at this point in time.

When we try to estimate those things, there is a single characteristic that seems to pop up every time, whether it's done by the Cost Analysis Improvement Group [CAIG] in OSD, or by the Congressional Budget Office, if you look at the assumptions, the cost estimate is as valid as the assumptions, but in the past, they have had one thing in common. Unfortunately, we've all been wrong on what the cost estimate has been or going to be.

So I think we need to work this problem really hard, and we will. As we approach deployments hopefully in the future we will have much sharper cost estimates. And as we approach our development program, we intend to have very specific cost estimates for our development efforts. We will obviously make them available to the Congress.

Mr. SHAYS. Mr. Christie or Dr. Stansberry, do you care to comment on that last question I asked?

Dr. Stansberry, let me just be clear on your role. You are the administrative side of this, in a sense. Mr. Christie is the testing side of this. Your role in missile defense, define it a little better for me, if you would.

Mr. STANSBERRY. My role in missile defense evolved out of the new oversight structure that Secretary Rumsfeld defined this year. And I am associated with the Missile Defense Support Group, which is independent of the Missile Defense Agency and evaluates the missile defense program, makes recommendations to the Under Secretary who is responsible for the administration.

Mr. SHAYS. So you have to be doing a lot of thinking on the very things we've been asking about.

Mr. STANSBERRY. Yes.

Mr. SHAYS. I don't want you to leave without making sure that you—I want to make sure that before you leave, you respond to any question that you have heard asked here. Is there any question you've heard asked that you want to put on the record, clarify, elaborate on or whatever?

Mr. STANSBERRY. Well, Mr. Chairman, I heard you raise a number of times questions about benchmarks. What we have done with the structure of the program is set apart a research, development, test and evaluation program that looks at all technologies for missile defense. Those are being run by the Missile Defense Agency, and General Kadish is responsible for that.

When that activity brings some element of a system to the point where it looks like it has miliary utility and PAC–3, the Patriot Advanced Capability Level 3 is at that point now, when some element gets mature enough that we want to procure it and deploy it, then it enters into a transition where it will move from General Kadish's organization into whatever is the appropriate service for the procurement and deployment.

While it's in General Kadish's organization, he is the authority for making acquisition milestone decisions. He has the authority for defining benchmarks. As individual elements move toward procurement and deployment, then they move into the standard acquisition process and they are then subject to acquisition decisions by the defense acquisition executive, advised by the defense acquisition board, with benchmarks that we are much more familiar with.

There will be benchmarks, there will be reporting to Congress. We have restructured the program for what I believe are very good reasons, and that has changed some elements of the benchmarks for those parts of the program that are in research and development.

Mr. SHAYS. I'm going to just comment that it strikes me that the benchmarks were dropped because they became very unrealistic and that we're making a determination to reexamine the benchmarks after we have a better handle on how workable this program is and when you can start to begin to even think about deployment. That's kind of what I'm hearing, and if I'm hearing wrong, I want to be corrected.

Mr. STANSBERRY. If you look back about 2 years, and compare the path we were on then, the path we're on today, I think you can understand some of the difference. We were 2 years ago under the ABM Treaty which required us to segregate very sharply defense against intercontinental range missiles from defense against shorter range missiles. We were on a path to deploy PAC–3, we were on a path to deploy what we called National Missile Defense.

With the result of pulling out of the ABM treaty, withdrawing from the ABM treaty, it has allowed us to combine technologies for defending against missiles of all ranges, where we couldn't before because of the treaty. That combining of technology led to us creating a single program, a single program for development. We backed off from plans to deploy the National Missile Defense, because it did not look like, once we got an integrated program, it did not look like that National Missile Defense architecture was going to be the right now.

So we don't know yet what will be the right architecture for defending the United States from long range ballistic missiles. That will depend to a large extent on the quality of the results out of the R&D program. And it's simply a fact of life that we can't specify those details right now, so we can't tell you what the deployed system would be, we can't tell you how much it would cost, we can't tell you when it will be out there.

Mr. SHAYS. Thank you.

Mr. Tierney, you have the floor.

Mr. TIERNEY. I'm astounded. Maybe somebody else gets this stuff, but I'll tell you, you're telling me, on the one hand, arroyo telling us there are benchmarks and on the other hand you're telling us there aren't. It seems that even within the research and development program there would have to be at some point some specific objectives on what you're trying to accomplish with your research and development, there would have to be some military requirements with respect to how you're going to get there, and there would have to be some time lines about when you think reasonably you ought to be there.

Now, that would be an ORD. And I haven't seen any reason or heard any explanation why you wouldn't still have ORDs applicable to this program, so that not just General Kadish in all his good wisdom and intentions would sit there some day and say, I think it's workable, it's on to the next stage. There's oversight committees in this Congress and elsewhere that ought to be able to look at that and say, general, what are your objectives specifically with this research and development program, what are your military requirements and what are your time lines. We want to take a look at it, I mean, we'll help you make that decision whether we want to keep chucking dough at this thing or not.

But just to say, oh, well, we don't want to do that any more because it got too complicated, so we pulled back, leads me to believe that you're going to keep testing this and testing it at $8 billion a year, and if it doesn't get to the point where you think it ought to get to the point, well, you'll just get another $8 billion and you go on, and if it doesn't work then you get another $8 billion and you go on. At no point do you chuck it up to the next level, where it says you want to go to deployment or whatever, and just keep on testing forever.

And Congress has no ability to say we should stop or we should go forward, because you haven't provided us with the objectives that you're trying to reach, the military requirements or the time tables for everybody to say, it's reasonable or it's not reasonable, you're within reasonable costs, we should continue on, you're within a reasonable pace of the program, you look like you're getting near the technology or not.

Abuse me of that notion if you would.

General KADISH. Congressman Tierney, I can assure you, we have those types of objectives. If you look at the airborne laser program to the ground based missile defense program, we certainly have the time line objectives and the testing objectives and the development objectives laid out, not only in our contracts, but in the plans we are developing.

It is true, however, that the guidance for those are no longer the operational requirements documents as we've known them, primarily because we need to see what we can do right now, rather than setting a bar that may not be reasonable, given the fact that we no longer have the ABM Treaty, for instance.

Mr. TIERNEY. At what point, or how is this committee at some point going to say to you, General, we think you've gone too far, we think you ought to stop, because you haven't reached this particular objective or some other? What do we measure by? How do we know if funds are being wasted?

General KADISH. I'm not sure, I'm at a loss for answering that question, except that we have an intense interaction with not only the Members but also the staff here in Congress.

Mr. TIERNEY. But you're taking away all our tools, General. Don't you get it? You're taking away the things by which we use to measure, and we used to be able to somehow determine it. That's what we had the ORD for, it was helpful for people like Mr. Christie, people like this committee and others to sit down there and say, OK, there are the objectives, we're looking at them, we can make a determination on the military requirements and the timeliness.

General KADISH. We will certainly share those. We have a document, for instance, called Technical Objectives and Goals [TOG], we certainly will share that with you.

Mr. TIERNEY. Well, let's talk about that for a second. You have described the so-called TOG, it's a much broader document, it gives much broader objectives but broader time lines than the previous document. How would you tell me that the TOG stacks up with the ORD?

General KADISH. The TOG, in my view, sets the stage for where we are today, an environment that we no longer have a demarcation between theater and national, we no longer have an ABM Treaty and we're trying to find the best deployable system within those guidelines together. Over time, I would expect those things to become more specific than they are today and actually be turned into capabilities assessment documents that military requirements will be specified in.

So this is a process, not a set of events. And I sympathize very strongly with your frustration of changing the way we do business, because——

Mr. TIERNEY. I don't mind changing business, I'm not glued to the past here. What I'm worried about is losing oversight capacity. When you tell me you've got this TOG business that's going to eventually get to more specific objectives and things like that, that doesn't do it for me. Because I've learned eventually, some things mean it's too far down the road, when you ought to have the information beforehand so you can make reasonable assessments.

Why can't you come to Congress with an ORD with respect to this program now? I have not heard yet, other than these broad talks with the GAO, we just don't think it's useful to do it that way. I would think that you would come in, even within just a research and development program, specific objectives and what you expect to accomplish within a specific time that you expect to get there, and show us what requirements have been met and have not been met. And why won't you do that? Why do you come up with this TOG instead, which is some broad thing down the road a bit, kicking the can down?

General KADISH. I believe that the approach we're using today is a much more flexible approach for the technology we're dealing with. Let me give you this example. The operational requirements documents that you specify have served this country very well. It's done by the military requirements community. It's a long, drawn-out process.

However, it's designed for mature technology. We write ORDs today for things like F–22, and I would specify that or stipulate that we, next year is 100 years of aviation manned flight anniversary. We are still writing operational requirements documents for airplanes, F–22, JSF, F and A–18s. They are pushing that state-of-the-art. That process is designed for mature technology, in my view, that we're pushing, pushing to its limits.

In the case of missile defense, we have a technology—in some of our unprecedented systems like SR–71, the Polaris program, the Apollo program—in those areas, we knew what our main objective was as four Presidents and Congresses since 1991 have been telling us, we want an effective missile defense. Now, how do we get it. I believe the processes we're putting in place today will handle your concern. But it will be different in the way we're trying to push this technology into fruition.

Mr. TIERNEY. Well, you're certainly not convincing me, General, I'll tell you. First of all, you're telling me now that you can't identify to Congress specific tests that you intend to conduct along the path here, and when those tests will be completed and whether or not——

General KADISH. I absolutely am saying we can do that, and I'll be happy to show you——

Mr. TIERNEY. But that's not part of your TOG, though, is it?

General KADISH. Maybe I'm not doing a good job of communicating that. But we have a test program laid out in great detail, certainly for the next 12 months and beyond. We will certainly give you that detail, whether it's classified or unclassified. And have, to the majority of the staff that have attended our briefings, we have it in great detail. The GAO, I believe, has at least eight looks going on in our program for various reasons as we speak. We are getting the detail for these types of things. So I——

Mr. TIERNEY. You're sitting there looking me straight in the eye telling me you have set out for us what specific tests you plan to conduct on which specific technologies by which particular date.

General KADISH. Yes. I look at that every week.

Mr. TIERNEY. And you have given that to Congress in some formal form?

General KADISH. Yes. And I have to go back and check, but our testing schedule is available to all those who need to know that.

Mr. TIERNEY. So the testing and evaluation master plan you produced to Congress, we just didn't know about it?

General KADISH. There is nothing, this formal document called the TEMP, we have for various and sundry of these elements. We have not reformulated it since January 2nd of this year, what we're going to use to go forward. But that doesn't mean we don't have specific test plans to go forward and——

Mr. TIERNEY. The TEMP was promised to us in June, now we're told we're not going to get it until October. Is October still the date we're holding for on the TEMP?

General KADISH. I'd have to go back and check. It's a rather laborious process to put that together.

Mr. TIERNEY. You're stunning me here a little bit, because this whole idea of having an ORD, which you're now telling me you can't do, you're telling me you've done everything, you just don't call it an ORD any more, that you have specific tests, that you're telling us what your objectives are, you're telling us what the military requirements are, that you're giving us some time lines, and you just renamed it, or you're not giving us those details?

General KADISH. We have much more flexibility in what we're doing today.

Mr. TIERNEY. That's what disturbs me a little bit here. Which part are you flexible? Are you not giving us specific objectives, you're being flexible there?

General KADISH. We have specific objectives in the near term plans that we have. What is different in terms of the process is that we can make more rapid decisions from an R&D point of view without having to go through the process called the ORD to change a requirement that particular document required. And from that standpoint, it gives us more flexibility to shorten our decision cycle times in order to get the development done.

I can give you some examples of what could happen in there, they are theoretical. But basically the time lines to get relief from a particular requirement, whether it is a big requirement or a little requirement, in the operational requirements documents, tended to be a lot more difficult than what we have structured today.

Mr. TIERNEY. What I find in this document, or rules that you have, some identification of when the time is that you expect you will have a system that is able to identify a decoy from a missile satisfactorily for you to say that you want to go to production.

General KADISH. I don't know if I can define an event quite that way. I think there will be a body of evidence that we will put together as a result of our test program, our analysis program, our modeling and simulation program that we will present and the weight of the evidence will either say, we can handle it or we can't. And we are working on that real hard right now. If I had to guess, it would be somewhere in the 2004 to 2006 range timeframe.

Mr. TIERNEY. But you're going to lay that out in some document, so that when it comes to that point in time, Congress will be able to take an assessment of where we are in relation to reaching that goal?

General KADISH. Our full effort is to get into that arrangement. It is a part of our technical objectives and goals to do just that.

Mr. SHAYS. We're just going to go for a few more minutes, then we'll be done.

Mr. TIERNEY. Mr. Christie, let me ask you, if we moved away from the traditional process of full range production and were to adopt this new incremental approach or whatever it is that the General is talking about there, what happens to the traditional role of evaluation and testing with you?

Mr. CHRISTIE. In this particular situation, where we're talking about capabilities-based acquisition, my role will be, or as I envision it right now, is giving time lines. At the end of each year, in fact, I will be providing to Congress my assessment of what's been going on.

Mr. TIERNEY. So you're going to get time lines from the General as to specific things that can be reached? Is that right?

Mr. CHRISTIE. Yes. For example, at the end of 2004, if somebody comes forward and says, we think we ought to take a hard look at where we are and whether we have a capability that might be worthwhile moving on into the next stage——

Mr. TIERNEY. You're not going to be looking at it in terms of, let's take a look at where we are compared to where we want to be, you're just going to take a look at where we are?

Mr. CHRISTIE. At where we are. I will assess what capability has been tested, what has been demonstrated in testing. And I will make my independent assessment of that.

That's different than the standard program, I admit, where you have an ORD and you have a specific set of requirements set out, which determines success or failure.

Mr. TIERNEY. So we're testing every kid in grammar school all the way through high school every damned year, but we're not going to test this program against any benchmarks at all, we're just going to occasionally look at it and see whether we want to keep slugging up the hill or not?

Mr. CHRISTIE. I will provide to the decisionmaker and to the Congress my independent assessment of what the capability is as we stand at that point in time——

Mr. TIERNEY. Based on what?

Mr. CHRISTIE. Based on the testing.

Mr. TIERNEY. You're going to tell us there's a Point A and Point B but you're not going to be able to tell us where they ought to be or it meets the benchmark or goal, or what——

Mr. CHRISTIE. That's not my job. I don't determine where we should be or not be.

Mr. TIERNEY. Well, you can determine whether or not they're successful as opposed to where they indicated they were going to be if they give you those kinds of measure, right?

Mr. CHRISTIE. If in fact I have a set of requirements.

Mr. TIERNEY. So now you won't, so you won't be able to do that, right?

Mr. CHRISTIE. I will be able to determine what capability we have based on the testing to that point.

Mr. TIERNEY. But you won't be able to measure it against anything, because you're not going to have any measurable objectives to look at.

Mr. CHRISTIE. If in fact we don't have a requirement, I will not be able to measure it against a requirement.

Mr. TIERNEY. Bingo. Right. Nor will anybody else.

General, let me just wrap this up with some questions here. You gave us a chart, or you had a chart in the beginning that was up there showing all the overall system, the four different elements of the boost phase segment, two different segments of the midcourse phase, four more elements of the terminal segment. And at the bottom, you factored in some communications and countermeasures and sensors and things. How much is this going to cost the American people?

General KADISH. I don't know.

Mr. TIERNEY. You don't know? Well, except for the fact you don't know, looking at the ground based, the midcourse system, it's essentially the same program that is now called an element of the larger system. So though we've had cost estimates of those before, so just with respect to that system, let's use the cost estimates, how much is that aspect of the system going to cost us?

General KADISH. Right now, we are spending somewhere, depending on how you measure it, $7.5 billion to $8 billion a year to research and develop those types of technologies. At some point in that process we will be able to decide whether or not we can put an architecture together to be effective.

So I can tell you that we're spending and costing the American taxpayer about 2 percent of our budget, a lot of money, but 2 percent of our budget——

Mr. TIERNEY. How much in hard dollars?

General KADISH. Pardon me?

Mr. TIERNEY. How much in hard dollars, as opposed to percentages?

General KADISH. About $7.5 billion a year.

Mr. TIERNEY. That's for the research.

General KADISH. That's for the research.

Mr. TIERNEY. Now, we've had these programs and the midcourse system and the cost estimates before, so using that, how much for the research and development issue, if we decide to go to development on this?

General KADISH. I'm not sure I understand your question.

Mr. TIERNEY. Well, we had, before we lumped all these things together, we had cost projections of what this ground based system was going to cost us. So I'm asking you, what is that system going to cost us, research and development through to completion?

General KADISH. At this point, I know what we will spend on it in terms of the budget we have programmed. But I don't know to completion. We're trying to define what complete means in this kind of an architecture. We have not reached that point yet.

So given that we haven't defined it, it's hard for us to estimate it. But there are some characteristics we could talk about. It will be expensive to continue. But in context of affordability, we're spending and programmed to spend about 2 percent of our DOD budget on this effort.

Mr. TIERNEY. How are we going to know, from the way that you've now designed this program, how is Congress going to know at some point that we're spending too much on this system, or too little, or just enough?

General KADISH. I think as we go through the authorization and appropriations process with the entire Congress that as we reveal our results every year, we'll either have great success and be able to move in an affordable way or great frustration, such that we'll be shut down. That's the normal way that programs work.

Mr. TIERNEY. No, normally they work because you have laid out some idea of specific goals and specific military requirements and specific time lines and we can measure against that. That's the way they normally work.

General KADISH. I can assure you, when we put the budget in front of Congress, we have a very specific set of details on what we're going to spend that money on, and we're held accountable for that. When the appropriations and authorization process review last year's results, what we plan on doing this year, and then the 5-year defense plan, that's the type of thing we've got before us.

Mr. TIERNEY. Thank you.

Mr. SHAYS. Mr. Kucinich, I will have two basic, unrelated questions to ask, and Mr. Kucinich just has a few.

Mr. KUCINICH. Isn't it true, General, that there are many operational and technical reasons why it's much more difficult to build a National Missile Defense System than to build an effective offense?

General KADISH. We've been building an effective offense for over 50 years. We've just begun to build a defense. So from that standpoint, we are at a disadvantage.

Mr. KUCINICH. Isn't it true that the attacker has a strong advantage because the defense must commit to a specific technology and architecture before the attacker does?

General KADISH. Can I take that one for the record? I'm not sure——

Mr. KUCINICH. OK, isn't it true that the defense will choose and then deploy hardware whose general characteristics will be known to the attacker?

General KADISH. The general characteristics, yes. But they would have to be very wary of the capability we bring to bear.

Mr. KUCINICH. So it's a good chance the offense is going to know the defense.

General KADISH. There's a good chance that the defense will know the offense.

Mr. KUCINICH. But it's more likely that the offense is going to know the defense.

General KADISH. In the military history, the way I read it, the supremacy of the defense and the offense have changed from time to time. In the Middle Ages, the defenses reigned and to get more modern times, the offense has reigned.

Mr. KUCINICH. Isn't it true the defense will not know with certainty what countermeasures the attacker would use?

General KADISH. And the attackers would not know with certainty what our capability is.

Mr. KUCINICH. You're giving us a lot of insight into the way you view this. It's very valuable, General.

I want to say that if we have very few parameters by which to judge success, and if there is no way for the Congress to effectively determine if you're over budget in any of these components, if there is no final goal to which the program is striving, and at the same time we find the role of Congress as far as oversight being undermined here, I'm just wondering, how does Congress perform its Constitutional duty to oversee this agency if there are no cost estimates, performance indicators or timelines available to us? You tell us, how do we do that?

General KADISH. Congressman, I believe that they are available. They're just not in the character of long term programs that we tend to be unrealistic in our objectives to try to meet. So I believe we are giving that kind of detail, and I'd be happy to come and spend some time with you and go over this, if that will help.

Mr. KUCINICH. I'll conclude, Mr. Chairman. I think, what I'd be happy is if you'd let the American people know more about what's going on here, instead of just individual Members of Congress.

I want to go back to one question. What does this cost? We've spent $70 billion as a Nation pursuing this fantasy of hitting a missile with a missile. What is this going to cost? We've already spent, Mr. Chairman, the taxpayers have spent $70 billion. Whether they're calling it Brilliant Pebbles or Star Wars or any of these other ideas, when is enough going to be enough, General? Should the taxpayer expect to give you a blank check? Is that what you're asking for?

General KADISH. I'm not asking for a blank check. I'm just trying to do what four Presidents and Congresses since 1991 tried to ask us to do.

Mr. KUCINICH. Thank you, Mr. Chairman.

Mr. SHAYS. Thank you.

I appreciate all three panelists. Obviously, General Kadish, we had more questions for you. So Dr. Stansberry, you'll probably go back and say, why was I there, except it was important that you be a resource for us. I thank you as well, Mr. Christie.

For me, I am going to evaluate this program on whether each test becomes more realistic. If we are not able to have a system that works, that's not going to be realistic, then I'm going to begin to wonder if we're going to be able to do what a number of us in Congress want to do. We haven't asked you about the laser, we haven't asked you about other options. So there's obviously things we could have discussed.

We will appreciate getting, for example, a listing of all the different variables that will ultimately tell us when we have a realistic system. And I would just quickly like to close with these two unrelated questions. I would like to know, what is the operational impact to missile defense of the delays in the spaced based infrared system load, the so-called Sabers Low program. What is the operational impact of this to the success of the program?

General KADISH. The Sabers Low intention was to have it as a space based sensor that would contribute to tracking and discrimination capability. At this point, we had always looked at the 2010, 2011 as being the point at which we might have an operational constellation. We have restructured that program, and right now we are looking to put early satellites into our test bed in the Pacific in the 2006, 2007 timeframe, if that's possible. That will help us decide what the future impact will be on Sabers Low.

Mr. SHAYS. Thank you. One last question. What upgrades are needed for the sea based program to have the system operational in an emergency situation, the launch of a ballistic missile against the United States? How long before these upgrades are completed?

General KADISH. The current standard missile three has very limited autonomous capability against long range missiles. It was not designed for that. We believe, however, there are early looks without treaty restrictions we are dealing with, that if we use offboard sensors, we may be able to improve that performance. But we don't know that yet. And the specific improvements would come out of that activity. We are looking at putting it in as a part of our program in the coming years. But we have not done so yet.

Mr. SHAYS. General, Mr. Christie, Dr. Stansberry, thank you very much for being here. Congress has mandated this program continue. Obviously it's going to have to make sure we constantly evaluate it. It is going to be interesting to see how we do it. I almost feel the analogy is research on cancer. We're doing research on a disease and we're not quite sure what it's going to cost us. Ultimately, we're hoping that we're going to have a system that works. I don't think anyone quite now can guarantee that it will work, frankly.

But I hope and pray we don't see a vote in Congress that actually votes to move forward with the program until we know it works. I have a feeling that you want to make sure the system works as well, before we deploy. So I thank you all very much.

Is there any question that we should have asked that you would like to put on the record, any point you want to put on the record before we leave?

General KADISH. No, Mr. Chairman.

Mr. SHAYS. Thank you very much for your patience.

[Recess.]

Mr. PUTNAM [assuming Chair]. The committee will reconvene for the purpose of seating panel two. Mr. Robert Levin, Director of Acquisition and Sourcing Management, U.S. GAO. He is accompanied by Barbara Haynes. Ambassador David Smith, the chief operating officer for the National Institute for Public Policy. Dr. William Graham, chairman and chief executive officer, National Security Research, Inc. And Mr. Eric Miller, senior defense investigator, the Project on Government Oversight.

As you know, we swear in witnesses in this subcommittee. So if you would please stand and raise your right hands.

Do you solemnly swear or affirm that the testimony you will give before this subcommittee will be the truth, the whole truth and nothing but the truth?

[Witnesses sworn.]

Mr. PUTNAM. Let the record reflect that the witnesses responded in the affirmative.

With that, we will recognize Mr. Levin for your opening statement. Welcome to the subcommittee.

STATEMENTS OF ROBERT E. LEVIN, DIRECTOR, ACQUISITION AND SOURCING MANAGEMENT, U.S. GENERAL ACCOUNTING OFFICE, ACCOMPANIED BY BARBARA H. HAYNES, ASSISTANT DIRECTOR; AMBASSADOR DAVID J. SMITH, CHIEF OPERATING OFFICER, NATIONAL INSTITUTE FOR PUBLIC POLICY; WILLIAM R. GRAHAM, FORMER DIRECTOR, OFFICE OF SCIENCE AND TECHNOLOGY POLICY, EXECUTIVE OFFICE OF THE PRESIDENT; CHAIRMAN AND CEO, NATIONAL SECURITY RESEARCH, INC.; AND ERIC MILLER, SENIOR DEFENSE INVESTIGATOR, THE PROJECT ON GOVERNMENT OVERSIGHT

Mr. LEVIN. Thank you, Mr. Chairman. I am pleased to be here today to discuss our review of the Airborne Laser. With me is Barbara Haynes, who is GAO's Assistant Director responsible for our missile defense work.

I think it's important to note how important the ABL is, the Airborne Laser, because the system, if successfully developed and fielded, will provide some capability to destroy enemy ballistic missiles, such as Scuds, soon after their launch. The ABL is also important because we are investing a lot of the taxpayers dollars in it. DOD has already received $1.7 billion for ABL development and projects that it will need another $2.7 billion for that development over the next 5 years.

My comments today are largely based on our report on the Airborne Laser that you requested, Mr. Chairman, and that has been released today. I will make five main points.

Point No. 1. The Missile Defense Agency faces huge technical challenges in developing the Airborne Laser. You saw that video earlier today of targets being intercepted. You would not be able to see an ABL system trying to intercept a target. It is not advanced enough for that yet.

When MDA took over responsibility for ABL from the Air Force last year, the program was 50 percent over budget and 4 years behind schedule. We found that the Air Force badly underestimated the complexity of the engineering tasks at hand and therefore misjudged the time and money that the program would need. Critical technologies that the system depends on remain immature. For example, the optics, the mirrors and windows that focus and control the laser beam and allow it to pass safely through the aircraft are at a technology readiness level of four. At TRL4, engineers have shown that the technology is feasible, but have not shown whether it will have the form, fit and function required in an operational system. This gets at the point about realism in testing.

MDA has a long way to go to get all the critical technologies to a TRL7, the level where it is demonstrated that components can work together as a system in a realistic environment. Our work over the years has shown that TRL7 is where it is appropriate to end technology development and begin system integration.

Point No. 2. You wanted us to comment on MDA's new strategy for developing the Airborne Laser. As you heard earlier, in January 2002, the Secretary of Defense directed MDA to quickly develop elements of the Ballistic Missile Defense System. In response to Secretary Rumsfeld's direction, MDA put in place a new requirement setting process. In the past, the military services would set requirements and systems developers would then try to build the technology to meet them. Under MDA's new strategy, the agency and the services now have the flexibility to match requirements with available technology. This is one of the knowledge based practices characterizing successful programs.

However, I want to emphasize that before the agency decides to go from technology development to system integration, it must freeze the requirements. At that point, there should be a match with available time, technology and money for developing and producing the system. The system developer can then begin designing a system, knowing full well what the requirements must be and what must be met.

Point No. 3. While MDA's new strategy incorporates much that is positive, we recommend in our report that the agency adopt another knowledge based practice. That practice is the establishment of decision points and associated criteria for separating the acquisition phases, the technology development, system integration, system demonstration, and production. Without such decision points and criteria in place, MDA risks beginning new and more costly activities before it knows how much time and money will be required to complete them and whether additional investment in those activities is really warranted, the kinds of points Mr. Tierney was raising.

Mr. Chairman, we were encouraged to learn last week that MDA is indeed considering how to fit such decisions points and criteria into the acquisition process for the Airborne Laser and other missile defense systems. They just haven't done it yet.

Point No. 4. You wanted us to comment on the role of the Operational Test and Evaluation office. Regarding OT&E's role, because MDA has not begun system level testing of the ABL system, operational test officials have had little involvement to date in overseeing test activities. However, they have begun working with MDA to construct a developmental test plan for missile defense overall. It hasn't been completed yet, as we heard earlier today.

At a later time, they expect to work on the test plans for each specific element, including the Airborne Laser. OTA has the authority to provide advice on that kind of test plan, but not to actually approve them.

Point No. 5, finally, you asked me to comment on changes and test plans in the absence of the Anti-Ballistic Missile Treaty restrictions. MDA can now test the ABL against longer range targets. Such testing would have been prohibited under the Treaty. MDA says, however, that it won't consider longer range targets until after it tests the ABL's ability to shoot down shorter range targets in 2004.

Mr. Chairman, in conclusion, I would like to stress that we think MDA has taken some positive actions, but we recommend that it adopt the practice of establishing decision points, and associated criteria, to separate the acquisition processes and phases. That ends my statement, and I would be happy to answer your questions.

[The prepared statement of Mr. Levin follows:]

United States General Accounting Office

GAO

Testimony

Before the Subcommittee on National Security, Veterans' Affairs, and International Relations, Committee on Government Reform, House of Representatives

For Release on Delivery
Expected at 10:00 a.m., EDT,
Tuesday, July 16, 2002

MISSILE DEFENSE

Knowledge-Based Process Would Benefit Airborne Laser Decision-Making

Statement of Robert E. Levin, Director, Acquisition and Sourcing Management

GAO

Accountability * Integrity * Reliability

GAO-02-949T

Mr. Chairman and Members of the Subcommittee:

Thank you for inviting me. I am pleased to be here today to discuss the Missile Defense Agency's strategy for acquiring ballistic missile defense systems. Pursuant to your request, Mr. Chairman, we issued a report on July 12, 2002, that examined whether the Missile Defense Agency's new strategy for developing the Airborne Laser includes practices that are characteristic of successful developmental programs.[1] Our observations today will reflect the knowledge that we gained from that work.

If the Department of Defense is successful in its efforts to develop and field the Airborne Laser, it could be an important system because it would give the United States some capability to destroy enemy missiles over the enemy's own territory. Enemy states that could launch ballistic missiles with chemical or biological warheads toward the United States, its deployed troops, or allies could be deterred from doing so if that enemy knew that the chemical and biological agents might fall back to earth over its own territory.

When the Air Force launched the Airborne Laser program in 1996, it estimated that developing the system would cost $2.5 billion and that it would be fielded by 2006. However, by August 2001, the Air Force determined that maturing the technologies and developing the system would cost $3.7 billion, or about 50 percent more, and the system would not be fielded until 2010, 4 years later than originally planned. Against this backdrop of cost increases and schedule delays, the Department of Defense, in October 2001, transferred responsibility for the Airborne Laser from the Air Force to the Ballistic Missile Defense Organization, now known as the Missile Defense Agency.

Because of your interest in the agency's recently expanded responsibility and authority for acquiring a capability to protect against enemy ballistic missiles, you requested that today we

- describe the Secretary of Defense's specific direction to the agency;
- explain the agency's requirements-setting process and plans for testing;
- provide details on the agency's investment decision-making process;

[1] U.S. General Accounting Office, *Missile Defense: Knowledge-Based Decision Making Needed To Reduce Risks in Developing Airborne Laser*, GAO-02-631 (Washington, D.C.: July 12, 2002)..

- explain how the maturity of technology critical to the system's design will be measured;
- comment on the role of the Director, Operational Test and Evaluation; and
- remark on changes in the agency's test plans since Antiballistic Missile Treaty restrictions are no longer in place.

To address our objectives, we reviewed documents and held discussions with officials at the Boeing Company, Seattle, Washington; Lockheed Martin, Sunnyvale, California; TRW, Los Angeles, California; Airborne Laser Program Office, Kirtland Air Force Base, New Mexico; Missile Defense Agency, Arlington, Virginia; and the Office of the Director, Operational Test and Evaluation, Arlington, Virginia. We conducted our review from August 2001 through July 2002 in accordance with generally accepted government auditing standards.

In summary Mr. Chairman, we found the following:

- In January 2002, the Secretary of Defense directed the Missile Defense Agency to quickly develop a ballistic missile defense system that included various elements, such as the Airborne Laser. These elements are to work together to defend the United States, its deployed troops, allies, and friends by engaging enemy ballistic missiles at various points during their flight. The Secretary also directed the agency to cancel existing requirements documents and, instead, develop elements with currently available technology that would be capable of defeating some, if not all, of the threat. This capability could then be improved over time.

- In response to this direction, the agency adopted changes that are characteristic of successful development programs. First, the agency adopted a flexible requirements-setting process that allows it to refine system requirements based on the results of system engineering. This process can result in less risk of cost and schedule growth because it does not establish requirements until systems engineering shows what is achievable. However, the agency must set the Airborne Laser's requirements when it determines that it has a match between the technology, money, and time needed to design and demonstrate an operational system so that the agency can use those requirements to measure the progress being made during product development. The agency is also implementing other changes to improve Airborne Laser's development. For example, the agency is allowing more time to mature and test technologies critical to the system's design, and it is improving ground test facilities so that in the future improved components can be tested on the ground before being installed in an aircraft.

- The agency has not yet implemented another practice that has proven beneficial in successful development programs and that we recommended to the agency in our July 12 report. Successful developers make decisions on whether to proceed with a program when the focus and cost of their activities are about to change. The criteria for beginning new activities is a determination by the developer that the program has the knowledge to begin the activities and that investing in them is the best use of the developer's resources. Our work over the years has found that the focus and cost of activities will change when the program is ready to move from technology development to system integration, then again into system demonstration, and finally on into production. Although, the Missile Defense Agency does not currently make decisions at each of these points or use the same criteria proven by successful developers, the agency is considering how to implement such a process.
- The Airborne Laser Program Office used technology readiness levels to assess the maturity of the technology critical to the design of the Airborne Laser. Our work shows that developers greatly enhanced their likelihood of success by beginning development only when they had demonstrated a system prototype in an operational environment. The program office's assessment showed that some Airborne Laser technology is almost to this point, but technology such as the mirrors and windows that focus and control the laser beam and allow it to pass safely through the aircraft (collectively referred to as optics) require additional engineering work before reaching this stage of maturity. Except for its evaluation of the laser, we agreed with the program office's technology assessment. We believe further testing is needed to demonstrate the maturity of the laser technology because the program has only tested a one-module laser (rather than the six-module laser planned for the first Airborne Laser configuration) in a controlled laboratory environment using surrogate components.
- By law, the Director of Operational Test and Evaluation (OT&E) is responsible for operational rather than developmental testing. However, OT&E officials said that being involved in developmental tests provides them insight and understanding to prepare for live-fire testing and later operational testing and evaluation. The officials told us that they have primarily been working on issues related to future live-fire lethality tests of the Airborne Laser, but they expect their

involvement in Airborne Laser's developmental tests will increase when system-level flight tests begin. The OT&E Director has not been given authority to approve Airborne Laser's developmental test and evaluation master plan, but OT&E officials have been asked to assist agency officials in preparing the plan and will provide comments before the plan is finalized.

- The Airborne Laser Program Office said that the United States' withdrawal from the Antiballistic Missile Treaty has not changed its plan to initially test the first Airborne Laser configuration against a short-range ballistic missile. However, the agency is considering future tests with longer-range missile targets. Such tests, which would have been restricted by the treaty, could occur in fiscal year 2005 or 2006.

Defense Secretary Directs New Approach for Acquiring and Deploying Missile Defenses

In 2001, the Department of Defense conducted missile defense reviews to determine how to best fulfill the nation's need to defend against enemy ballistic missile attacks. As a result of these studies, the Secretary of Defense in January 2002 directed that the Ballistic Missile Defense Organization be elevated to the status of an agency and renamed as the Missile Defense Agency. The Secretary's key priorities were to field quickly an integrated missile defense system that could defend the United States, deployed forces, allies, and friends by engaging enemy ballistic missiles at various points during their flight. Another of the Secretary's priorities was to provide an early capability by using test assets or prototypes in the event of an emergency. To enable the agency to achieve his priorities, the Secretary directed the agency to abandon its traditional requirements-setting process that required a military service to establish technical requirements when a weapon system acquisition program was launched and adopt a more flexible, capability-based process that would allow the agency to use available technology to develop a weapon system that could engage some, if not all, of the current threat. This "base-line" capability would then be improved over time. The Airborne Laser system was one of many systems affected by these changes.

Agency Adopts New Requirements-Setting Process and Other Practices

In response to the Secretary's direction, the Missile Defense Agency adopted a new Airborne Laser development strategy that incorporates some of the practices characteristic of successful programs. These practices include a more flexible requirements-setting process, allowing more time to mature and test the Airborne Laser's critical technologies, and improving test facilities.

Our work shows that the flexible requirements-setting process can result in less risk of cost and schedule growth because requirements are not set until systems engineering shows what is achievable. However, the agency must set the Airborne Laser's requirements once it determines that it has a match between the technology, money, and time needed to design and demonstrate an operational system so that the agency can use those requirements to measure the progress being made during product development.

The Department of Defense ordinarily faces significant hurdles in matching requirements to available resources (time, technology, and money). The fundamental problem is two-fold. First, under the Department's traditional process, requirements must be set before a program can be approved and a program must be approved before the product developer conducts systems engineering. Second, the competition for funding encourages requirements that will make the desired weapon system stand out from others. Consequently, many of the Department's product development programs include unrealistic requirements set by the user before the product developer has conducted the system engineering necessary to identify the time, technology, and money necessary to develop a product capable of meeting requirements.

The agency is also adopting other practices that are likely to improve Airborne Laser's development. These practices include allowing more time for testing and developing facilities to mature and test critical technologies. The agency is initially developing and testing a six-module laser system to demonstrate technologies critical to the Airborne Laser's design. When the Air Force was responsible for the Airborne Laser program, it planned to complete system-level flight tests of the six-module Airborne Laser system in the last quarter of fiscal year 2003, but the agency has delayed completion of the test to the first quarter of fiscal year 2005. This delay allows additional time to learn from and correct problems discovered during the tests. In addition, the agency plans to increase the Airborne Laser's ground-testing capability by awarding a contract in 2003 for what they are calling an "iron bird," which is essentially an aircraft hull with laser equipment installed. The "iron bird" is expected to allow testing of a fully integrated Airborne Laser system on the ground so that technologies for future configurations can be evaluated before being installed in an aircraft.

Allowing more time for testing is important because testing informs the requirements process. Because testing allows developers to gauge the progress being made in translating an idea into a weapon system, it enables the developer to make a more informed decision as to whether a technology is ready to be incorporated into a system's design. With this knowledge, the developer can determine whether the technology is so important to the system's design that additional time and money should be spent to mature the technology or whether the system's initial performance requirements should be reduced.

The "iron bird" is expected to reduce the cost of testing technologies planned for future Airborne Laser configurations. With it, the agency can mature new component-level technologies to higher levels in the less expensive ground-testing environment before installing them on an aircraft.

Changes in Agency's Decision-Making Process Are Needed

We reported in July 2002, that the Missile Defense Agency's new Airborne Laser acquisition strategy does not include decision points with appropriate knowledge-based criteria for moving the Airborne Laser program forward. However, the agency is now considering how it can implement such a process.

In successful developments, developers make decisions when the focus and cost of program activities is about to change. At these points, the developers decide whether they have the knowledge to begin new activities and whether investing further time and money in their product is the best use of their resources. The first decision point occurs when the focus of the developer's activities change from technology development to system integration. This point is reached when the developer has incorporated technology into subcomponents with the form, fit, and function needed in an operational system and the developer is ready to design a system that integrates those subcomponents. The criterion for deciding to move the program forward is having the knowledge to match requirements and available resources (time, technology, and funds). The second decision point occurs between system integration and system demonstration when the developer has successfully integrated subsystems and components into a design that not only meets the customer's performance requirements but also has optimized the design for reproducibility, maintainability, and reliability. A developer moves the program forward at this point only if the design is stable, which is generally considered to be the point at which about 90 percent of the design's engineering drawings have been released. The third decision point separates system demonstration from production. The decision to invest in production is based on having the knowledge to determine that the product performs as required during testing and that the manufacturing processes will produce a product within cost, schedule, and quality targets.

Figure 2: The Knowledge-Based Process

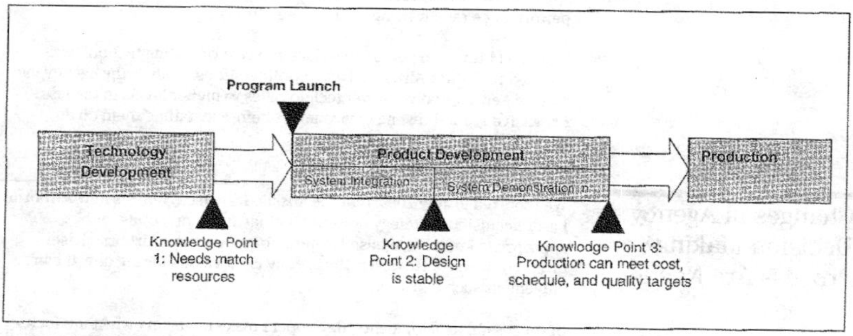

Decisions are made at these points not only because the focus of activities is changing, but also because the cost of the activities are increasing. Our work shows that product development is typically much more costly than technology development. This is because during technology development, small teams of technologists work to perfect the application of scientific knowledge to a practical problem. As product development begins, developers begin to make larger investments in human capital, bringing on a large engineering force to design and manufacture the product. In addition, product development requires significant investments in facilities and materials. These investments increase continuously as the product approaches the point of manufacture.

The Airborne Laser acquisition process has three phases separated by two decision points. The phases are development, transition, and procurement. The development phase includes all developmental activities and system level demonstrations of military utility. Transition includes preparation of the operational requirements document by the appropriate armed service and operational testing; and production includes producing and fielding the weapon system. The first decision point occurs between development and transition. A decision will be made to begin the transition phase if the agency determines that it has the technology in-hand to produce a system that merits fielding. At the end of the transition phase, the Airborne Laser would enter the formal Department of Defense acquisition process at Milestone C—the point at which the Department decides whether a system should begin low-rate initial production.

The Missile Defense Agency's current decision-making process puts at risk the agency's ability to develop a useful military capability on time and within budget. This is because the agency's process does not include an established set of decision points with appropriate knowledge-based criteria for deciding whether to invest in system integration and, subsequently, system demonstration and production. For example, the agency does not separate technology development from system integration with a decision point or use the knowledge the program has attained to determine if the technology can be incorporated into a mature system design within available time and funding constraints. Agency officials are considering how to fit such decision points into Airborne Laser's acquisition process, as well as the acquisition process of other elements.

Technology Readiness Levels Are Used to Measure Technology Maturity

In 2002, the Airborne Laser Program Office assessed the maturity of technologies critical to the development of the Airborne Laser system. To make this assessment, the program office used a tool known as technology readiness levels that was developed by the National Aeronautical and Space Administration. A comparison of a 1999 assessment and the 2002 assessment shows that the Airborne Laser program has made progress in maturing critical technologies, but much remains to be done.

In 1996, the Air Force launched the Airborne Laser program to develop a defensive system that could destroy enemy missiles from a distance of several hundred kilometers. Engineers concluded that if they were to meet this requirement, the system would need a fourteen-module oxygen iodine laser. They also determined that the system would need a beam control/fire control assembly that could (1) safely move the laser beam through the aircraft, (2) shape the beam so that it would not be scattered or weakened by the atmosphere, and (3) hold the beam on target despite the movement of the aircraft. In addition, engineers determined that the system would need a battle management and control system capable of planning and executing an engagement. To determine if the technology was "in hand" to meet this requirement, the Air Force planned to build a six-module Airborne Laser configuration and test it against a short-range ballistic missile. Under the Missile Defense Agency's new development strategy, this six-module configuration is now known as Block 2004.[2]

Officials began their 2002 assessment by determining the technologies critical to designing Block 2004 and future configurations. These technologies are: (1) devices that stabilize the laser system aboard the aircraft so that the beam can be maintained firmly on the target; (2) optics—mirrors and windows—that focus and control the laser beam and allow it to pass safely through the aircraft; (3) optical coatings that enhance the optics' ability to pass laser energy through the system and to reflect the laser energy; (4) hardware that works in tandem with computer software to actively track the target missile; (5) devices that measure atmospheric turbulence and compensate for it so that it does not scatter or weaken the laser beam; and (6) safety systems that automatically shut down the high energy laser in the event of an emergency. At our request, the Airborne Laser Program Office also assessed the maturity of the oxygen iodine laser.

The program office assessed the optics and stabilizing devices at technology readiness level four, the optical coatings at level five, and the safety systems, atmospheric compensation, and target tracking devices at level six. At level four, scientists have shown that a technology is technically feasible, but have not shown whether the technology will have the form, fit, or function required in the operational system. When a technology progresses to level five, the technology being tested is incorporated into hardware whose form and fit are coming closer to that needed for an operational component and that hardware is integrated with reasonably realistic supporting elements so that the technology can be tested in a simulated environment. Finally, at technology readiness level six, integrated testing of a prototype system has occurred in a relevant environment. Our prior work has shown that demonstrating a technology readiness level of seven, that is, demonstrating that components can work together as a system in a realistic setting, prior to establishing cost and schedule estimates and beginning system integration, is an important determinant of program success.

[2] This configuration is known as Block 2004 because it is to be completed in December 2004.

We disagreed with the agency's assessment in only one instance—the assessment of the maturity of the system's laser component. The agency assessed the maturity of the oxygen iodine laser at a readiness level of six while we consider it to be at a level four. The tests of a one-module laser that the program office relied upon to prove the maturity of the laser were conducted in a controlled laboratory environment with surrogate components. In our opinion, the program office will demonstrate the laser technology in a relative environment (technology readiness level six) when the six-module system is integrated and successfully tested at full power within the high fidelity laboratory environment of the Airborne Laser Systems Integration Laboratory, currently under construction at Edwards Air Force Base, California.

Operational Test Community Expects to be Involved in Developmental and Operational Testing

By law, the Director of the Office of Operational Test and Evaluation (OT&E) is the principal advisor to the Secretary of Defense and the Under Secretary of Defense for Acquisition, Technology, and Logistics on operational test and evaluation of Department of Defense weapon systems and is also responsible for monitoring and reviewing live fire testing activities.[3] The Director may not be assigned any responsibility for developmental test and evaluation, other than to provide advice to officials responsible for such testing. However, OT&E officials said that being involved in developmental tests aids them in preparing for live fire testing and later operational testing and evaluation, for which they are responsible.

The officials told us that they have primarily been working on issues related to future live-fire lethality tests of the Airborne Laser, but they expect their involvement in Airborne Laser's developmental tests will increase when system-level flight tests begin. This expectation is based on the agency's request that the OT&E officials work with Missile Defense Agency officials to construct developmental test and evaluation master plans for the integrated missile defense system and each element. Although OT&E has not been given authority to approve the Airborne Laser's developmental test and evaluation master plan, OT&E officials said that they will assist in preparing the plan and will review and offer comments before it is finalized.

[3] 10 USC 139

In addition to participating in the creation of the developmental test and evaluation master plans, the Director, OT&E, will participate in approving each element's operational test plan and will evaluate their operational tests. Further, under the agency's new strategy, OT&E officials are members of the Missile Defense Support Group and its working group. These groups are expected to provide insight and advice to the Missile Defense Agency Director and to the Senior Executive Council that is responsible for all major missile defense decisions. The agency has also assured the operational test community that it will be included in all meetings and reviews regarding testing so that test officials can annually review and report on the adequacy and sufficiency of the Missile Defense Agency's testing program.

Targets in Future Tests Could Change

Airborne Laser program officials told us that the Airborne Laser was originally planned as defense against short-range ballistic missiles. However, with the United States withdrawal from the Antiballistic Missile Treaty, the Missile Defense Agency is considering tests against longer-range missiles. While such a change would not affect the program office's plan to test the Airborne Laser against a short-range missile in December 2004, tests conducted in fiscal year 2005 or 2006 could include targets representative of longer-range missiles. The Antiballistic Missile Treaty would have restricted tests that include longer-range missiles as Airborne Laser targets.

Conclusion

In an effort to field a missile defense capability quickly, the Department of Defense has directed the Missile Defense Agency to adopt a new acquisition strategy. Some of the practices that we observed are being implemented in the Airborne Laser program are practices that have been proven in successful development programs. For example, similar to successful developers, the agency has deferred the establishment of the Airborne Laser's requirements until the knowledge is available to set realistic requirements. Successful developers also improve their product incrementally, just as the agency plans to improve the Airborne Laser through a series of upgrades. However, the agency has the opportunity to make its acquisition process more knowledge-based. By establishing knowledge-based decision points at key junctures, the agency would be in a better position to decide whether to invest in the next phase of the Airborne Laser's development. Also, the agency would be better able to hold the Airborne Laser Program Office accountable for planning all of the activities required to develop a quality product, approaching those activities in a systematic manner so that no important steps are skipped and problems are resolved sooner rather than later, and making cost and schedule projections when they have the knowledge to make realistic estimates. With this disciplined process in place, the agency will be much better positioned to decide whether to invest further in the Airborne Laser or use available time and funds for some other element of the missile defense system.

Our July 12 report recommended that the Director of the Missile Defense Agency not only establish decision points to separate technology development from system integration; system integration from system demonstration; and system demonstration from production, but also establish knowledge-based criteria that would be used to determine whether additional investments should be made in the Airborne Laser program. We are encouraged that the agency is now considering actions that could prove to be significant steps toward implementing these recommendations.

- - - - -

Mr. Chairman, this concludes my statement. I would be pleased to respond to any questions you or members of the committee may have.

Mr. PUTNAM. Thank you very much.

At this time, we will take testimony from Ambassador David Smith, the COO of the National Institute for Public Policy. Welcome.

Ambassador SMITH. Thank you, Mr. Chairman. It's a pleasure to be here and to see you again, sir. It's an honor to appear before this committee, speaking on behalf of President Bush's missile defense program, to defend America and its overseas forces, its allies and its friends.

I should stress that this program really is very much a part of homeland defense. On September 10th, I would have argued that missile defense was absolutely essential, because unchecked, ballistic missiles would prove alluring to potential adversaries, undermine non-proliferation, confer asymmetric geopolitical advantages, and eventually develop into combat capabilities that could threaten U.S. military mission success.

As a former intelligence officer, I can tell you that if we look at the situation it's not like rejecting the old Soviet threat. It's not that neat. We understood the Soviets a lot better. So we don't have that kind of thing.

Do we know who our potential adversaries are and what they're doing? You bet we do. The trend is toward more missiles in the hands of more countries with greater range, greater accuracy, multiple launch modes, and a variety of specialized weapons.

What General Kadish has to do is essentially plan against what is essentially a rolling composite. That doesn't mean he has no information or that he does not know what he is planning against. It means that it's just a heck of a lot harder to do than it was against the old Soviet Union.

Can we defend against ballistic missiles? You bet we can. When was the first time an ICBM was shot down by a defense interceptor? In 1962, Nike X shot down an Atlas ICBM. Now, you've seen in General Kadish's video another few of the more recent successes. Am I surprised that we're walking before we run? Absolutely not. When I was in military service, they took me out to a rifle range and told me to fire an M–16 at a stationary target before they let me out anywhere else with that rifle. That is exactly how you're going to do that.

If you were any more successful, frankly, you'd kind of wonder what he was up to. Of course there are going to be a few so-called failures. There's really no such thing as a failed test.

Now, tragically, everything I have said about why we need to do missile defense was also true on September 12th. Yes, Al Qaeda has been dominating the headlines. They employed a particular kind of asymmetric strategy, which involved turning an ordinary item, an airliner, into a cruise missile carrying weapons of mass destruction. But I assure you that other adversaries will also use asymmetric strategies, and they will use different means. Ballistic missiles remain attractive to many.

I'd like to quote Senator Joe Lieberman, who spoke 2 days after September 11th, saying, "September 11th certainly shows we're vulnerable to more than just missile attack. But we're vulnerable to missiles, too." That strikes me as eminent common sense.

The choice of which threat warrants our investment and which one we can ignore is simply not ours. We must not delude ourselves. Will this be costly? Yes, it will. But as Senator Arlen Specter remarked also a few days after September 11th, we do have the resources to do both. That is exactly what President Bush has set out to do.

Now, one important step was withdrawing from the ABM Treaty. The fact is, the cold war ABM Treaty prohibited any kind of defense of the territory of the United States, any kind, including a single site in North Dakota. It precluded development and testing of anything but a fixed land based system, and the endless diplomatic hashing of this with the Russians simply preserved mutual assured destruction long beyond its due.

Now we need to proceed with a normal testing pattern. We're already butting up against it, there's nothing that has to be done artificially. Something as simple as turning on the SPY–1 radar on an Aegis ship during a missile defense test was precluded by the ABM Treaty. We are now able to do that.

Given President Bush's objectives in getting out of the ABM Treaty, the Secretary of Defense's guidance to the Missile Defense Agency seems appropriate. First, a limited defense of the United States in the near term and the defense of our overseas forces, allies and friends. A layered defense, ready as soon as practicable, and an early capability if necessary. This restores three essential points that had been missing from our missile defense program for the earlier 8 years, globality, layering and evolution. Evolution is absolutely essential to get out and stay out ahead of the threat. It is the only way we are going to keep our potential adversaries from acquiring missiles, is to tell them that they cannot succeed. It is also essential to adapt to a very complex evolution of the threat.

When something happens and we have to ask General Kadish what it is that he has, what is he capable of doing tomorrow morning, the American people will want some kind of a capability rushed into that situation, much as we did with the not quite ready for prime time PAC–2 during the Gulf war. That is the measure of success, is against the capability, it's the capability against the need. The American people will not be looking for an ORD which matures 2 years later.

This kind of evolution requires the capability based approach that General Kadish describes. How many calls for change have we had in our acquisition system? How many in this body have decried that it takes two decades from concept to field a military system?

I think this can all be done with congressional oversight and DOD oversight, as General Kadish and Dr. Stansberry and Mr. Christie outlined earlier, a system for oversight is absolutely possible. That effective oversight extends to the subject countermeasures that DOD has, in my view, appropriately classified. We do not want to discuss what we know and what we do not know about countermeasures. We do not want to discuss what we have tested. We do not want to discuss vulnerabilities. We would not ask, for instance, the director of the F–22 program or the joint strike fighter or any of our programs to come and talk about the vulnerabilities of those systems, even in their earliest stages, because that gives an awful lot of information out there. This information is available to the Congress if the Congress wants it, in closed session with appropriate safeguards.

And the fact is, one of those systems that General Kadish is talking about may just have to be out there a little sooner than anyone expected. He is charged with providing an early capability if necessary. No one expected PAC–2 to be in the Middle East when it was, and there it was, and it did its job to the best of its ability at that time.

Mr. Chairman, thank you very much. That sums up what I had in my written testimony and I would be pleased to take any questions or any discussion my remarks have prompted. Thank you.

[The prepared statement of Ambassador Smith follows:]

Remarks of Ambassador David J. Smith
before the
House of Representatives
Committee on Government Reform
Subcommittee on National Security, Veterans Affairs, and International Relations
July 16, 2002

Mr. Chairman:

As a former House and Senate staffer it is always a pleasure to return to Capitol Hill. And it is indeed an honor to appear before this subcommittee in support of President Bush's missile defense program for America, its overseas forces, allies and friends.

You have asked me to address:

- the relation of missile defense priorities to homeland defense;
- how these priorities should be set; and
- what testing should be undertaken in the post ABM Treaty environment?

I shall address these topics in the context of post Cold War, post September 11 national security strategy, as I see it. I should stress that these are my personal views; they do not necessarily represent those of the organization that I serve.

Before September 11, I frequently argued that missile defense was essential because, unchecked, ballistic missiles will continue to prove alluring to many potential adversaries, undermining nonproliferation efforts, conferring asymmetric geopolitical advantages, and eventually developing into combat capabilities that could threaten US military mission success.

That said, in the aftermath of September 11 and the national security requirements the horrific events of that day generated, it was appropriate to consider the relative priority of missile defense. My analysis indicates that the priority President Bush has steadfastly assigned it remains valid. Consider the following observations:

- Because the US has enduring worldwide interests and commitments, we will remain engaged economically, diplomatically and militarily.
- Consequently, we will have serious enemies.
- Although Al Qaeda currently dominates the headlines, it is hardly the only enemy or potential enemy. September 11 surely does not signal that henceforward only thuggish cave dwellers with twisted concepts of religion will direct history. Post Cold War potential adversaries continue to include sub and transnational groups, quasi states, mini states and emerging mid size powers.
- Some serious adversaries will go to extraordinary lengths to achieve their objectives.
- America, and the entire free world, is vulnerable because of the very nature of its society.

- As a society, we are slow to prepare defenses, even in the face of repeated warnings.
- The end of the Cold War coincided with a sharp growth in the availability of sophisticated technologies with military and terrorist applications.
- For the foreseeable future, any attempt to take on the US and its allies will be asymmetric.

Turning commonplace airliners into cruise missiles carrying weapons of mass destruction in the form of jet fuel was a quintessential asymmetric strategy. But potential adversaries with differing objectives and differing means will employ different asymmetric strategies. And, for some, the ballistic missile remains a perfect asymmetric strategy. September 11, argued Senator Joseph Lieberman just two days later, "certainly shows we're vulnerable to more than just missile attack, but we're vulnerable to missiles too."

To be sure, we must now prepare for terrorism petty and grand. But the ballistic missile threat to the United States, its armed forces and its interests is no less today than on September 10. We must not delude ourselves! Although how we respond will have some effect on our adversaries' choice of means, the choice of which national security threat warrants our investment and which to ignore is not ours. We must protect against terrorism and proceed apace with missile defense. Moreover, argued Senator Arlen Specter, "we have the resources to do both."

This is exactly what the Bush Administration has set out to do. While prosecuting the global war on terrorism, they issued the Quadrennial Defense Review on September 30, 2001, gave notice of US withdrawal from the Cold War ABM Treaty on December 13, 2001 and completed the landmark Nuclear Posture Review (NPR) on January 9 of this year. The ABM Treaty and the NPR warrant a brief comment in the context of this hearing.

A decade after the Cold War's end, it was clear that the 1972 ABM Treaty was burdensome to US defense and to US-Russia relations. The ABM Treaty simply prohibited any missile defense of United States territory, whether from a single site in North Dakota or anywhere else. Moreover, it prohibited the development and testing of all but fixed, ground based interceptors. The vacuum thus created spawned a sterile meta debate over whether missile defense would work. Of course it would -- if only Yankee ingenuity were allowed to work on it! And the constant rehashing of the matter at every US-Russia diplomatic meeting simply kept Cold War mutual assured destruction on artificial life support. It was time to defend America and forge a new and better relationship with Moscow. The ABM Treaty slipped away on June 13 with nary a whimper. Despite all the shouting and predictions of doom, it just didn't matter!

There is no need for an artificial rush past the ABM Treaty's limits. Rather, the Missile Defense Agency (MDA) should simply proceed with the rational development and testing program it has laid out. Undramatic steps such as connecting an active Aegis Spy-1 radar to the rest of the command, control and communications system during a missile defense test will serve the program's purpose and, by the way, soon underscore that we were already butting up against the ABM Treaty's limits. Testing unfettered by antiquated restrictions is the quickest path to the President's goal of missile defense for America, its overseas forces, allies and friends. And when it comes to deployment, we can now consider the most effective near term solutions such as a combination of multiple ground-based sites, seaborne and airborne defenses. In the medium term, effective missile defense must include one or more space based elements, and we are now free to develop and test these options.

This emphasis on missile defense is placed in a sound post Cold War strategic context by the NPR. Henceforward, the US will rely upon a new triad: considerably reduced traditional strategic offensive forces coupled with strategic conventional forces, a responsive infrastructure to prepare for uncertainty, and increased reliance on defenses. Reductions in strategic nuclear weapons have been placed in the context of a new strategy that is better suited to the future. This is the best opportunity for further significant reductions in strategic offensive forces because, so long as mutual assured destruction prevailed, there was, frankly, a practical limit on the extent of reductions. The NPR provides a new strategy and fills in the needed new systems, allowing further reduction of Cold War strategic offensive forces.

Given the President's missile defense objective and the national strategy outlined in the NPR, the goals assigned by the Secretary of Defense to the newly established MDA are altogether appropriate:

- defend the United States against limited attack and defend its overseas forces, allies and friends;
- employ a layered system;
- enable deployment as soon as practicable; and
- provide an early capability, if necessary.

These four points restore three essential elements that had been missing from US missile defense policy for eight years: globality, layering and evolution.

The march of ballistic missile technology -- more missiles in more countries, greater range, greater accuracy, multiple launch modes and more specialized warheads -- as well as the events of September 11 underscore the requirement for direct defense of our homeland. At the same time, we must continue our traditional policy of forward defense, not just to defend our worldwide interests, but as an effective strategy to defend our homeland. Furthermore, MDA Director Lieutenant General Kadish should be commended for pointing out that a short-range missile launched from a forward position could threaten the United States itself. Consequently, only a seamless missile defense system that eschews the artificial, ABM Treaty based distinction between NMD and TMD makes sense.

Globality and effectiveness require a layered system. No single element of a missile defense system -- or any system, for that matter -- can be completely effective. It is high time to squelch the rhetoric that holds missile defense to an absurd, celestial standard, and recognize that a layered system that maximizes potential shots and leverages the strengths of each element at the right moment can and will be very effective -- not perfect; very effective.

And our system must be evolutionary. We should be prepared to cobble together an early capability, if necessary, aim toward an achievable system in a few years, and make clear that we will get out ahead, and stay out ahead, of the developing threat as long as necessary. Potential adversaries of every description must know that they confront America's insurmountable technological prowess. Getting out ahead of the evolving threat is the only way to affect it and, if necessary, to defend against it. Otherwise, we can be sure that potential adversaries will continue to seek the leverage that ballistic missiles can bestow.

Evolution in an uncertain world requires the much-maligned capabilities based approach. As a former intelligence officer, I can tell you that the Soviet Union was relatively easy to follow. We understood their development, testing and acquisition cycle and only occasionally scrambled to understand a surprise breakthrough or unanticipated failure. Despite the secrecy, the Soviets were relatively predictable. As an analyst who tracks worldwide ballistic missile developments, and a consultant to the Rumsfeld Commission, I can tell you that today we do not have the analytical luxury of yesteryear. The facts as we know them -- and also what we do not know -- hold the potential for a number of different outcomes. A capabilities based approach provides constant reevaluation of a composite threat, the only approach that can support a design and acquisition process that emulates our most innovative civilian industries.

The flexibility that the capabilities based approach affords must be matched by the DoD acquisition system. How many calls have there been over decades to reform the Pentagon's acquisition system? How many Members of Congress have decried that it takes two decades to field a US military system? Well, DoD and the MDA are pioneering a fix. It will no doubt need fine-tuning over the coming years, but missile defense requires that we do something to shorten the concept to field loop.

This new way of proceeding from design, through research, development and testing, to deployment can be achieved within the bounds of effective oversight. General Kadish has detailed how the system insures the involvement of the Secretary of Defense, the Deputy Secretary and the Under Secretary for Acquisition, Technology and Logistics.

This new flexibility in the Pentagon will also require some flexibility on Capitol Hill. As I understand it, the format and timing of some of the information sent to the Congress will change, but the essentials of Congressional oversight are preserved. Moreover, DoD will surely answer any Congressional questions and clarify any uncertainties that the transition to a more streamlined acquisition process will entail. What is important is that the Executive and Legislative branches work together to develop a streamlined missile defense acquisition process, perhaps one that could be extended to other areas of the Pentagon.

With regard to Congressional oversight, there is one more issue I would like to address. The issue of countermeasures is serious and I believe US missile defense designers are treating it accordingly. But serious treatment of a threat and how we design against it also require circumspection with regard to the information available on the subject. Unfortunately, countermeasures have become the latest chapter in what I earlier called the sterile meta debate over whether missile defense will work. Ironically, many of the same analysts who just a few years ago scoffed at the very idea of any missile threat, and that we could "hit a bullet with a bullet" (which we have now demonstrated we can do), now regale us with tales of countermeasures -- some plausible and some fantastic -- sometimes providing potential adversaries with fruitful avenues of exploration, sometimes suggesting weaknesses in design of US defense systems.

This public free-for-all is counterproductive. As a citizen and taxpayer I am glad that DoD has decided to classify its real threat analysis and the counter countermeasures work it is doing. There are plenty of avenues for independent scientists to assist the Pentagon. And classification is no bar to effective Congressional oversight. Our elected representatives have access to classified details of our military programs, including special access programs that are entirely classified. Good stewardship of the country's business requires accounting for every dollar, ensuring appropriate priorities and insisting upon sound program management. But it surely also requires safeguarding sensitive information.

Mr. Chairman, I believe I have addressed each topic posed to me. I hope that the context in which I have placed my analysis is helpful to you and this subcommittee. I would welcome any questions or discussion my remarks may have prompted.

Thank you.

Mr. PUTNAM. Thank you, Mr. Ambassador.

At this time we recognize Dr. William Graham, chairman and chief executive officer of the National Security Research, Inc. Welcome.

Mr. GRAHAM. Thank you, Mr. Chairman, distinguished members of the subcommittee, and thank you for inviting me to testify on my views concerning the challenges and the opportunities that are now available in the missile defense area to enhance our national security and our homeland defense.

I am testifying today on my own behalf, and not as a representative of any organization. However, I would like to state for the record that after I submitted my prepared statement, I determined that National Security Research, the company for which I am CEO and chairman, does some work in the area of missile defense and has a current subcontract that provides advice to a prime contractor in analyzing ballistic missile threat characteristics.

The prospective threat to our security from long range ballistic missiles, such as the North Korean Taepo-dong 2, is generally understood. The United States is developing missile defenses to deal with such a threat. However, this alone would be akin to locking the front door and assuming the house would be safe. The United States, in fact, is also vulnerable to attack from short range missiles. This might seem strange, because we're a continent isolated by two large oceans. But because this vulnerability has not received the same attention of long range ballistic missiles, I would like to spend a moment today addressing that threat. And I will characterize the threat in my statement and we can perhaps talk about approaches to dealing with it in the question and answer, if you wish. There are discussions of those approaches in my testimony.

But we also need to make sure the back door is locked before we can adequately ensure the safety of the homeland against ballistic missile threats. One of the worst things we can do is leave ourselves completely undefended from any threat, because that in fact is provocative and an invitation for others to use that as a means to attack us.

The termination of the ABM Treaty last month has finally opened the opportunity to develop effective missile defense capabilities for those of us who have waited for the time to come when a full range of engineering options and missile defense could be explored. And I have testified in Congress for about 10 years on the need for terminating that Treaty. So my theme here is quite consistent.

Gone are the constraints, finally, in the United States that were drafted for the security environment of the 1960's and 1970's. This freedom permits the United States to prepare for threats to our security from most traditional adversaries that may threaten us far away and now to prepare for threats that are more ambiguous and possibly within a short range of our soil, as I will describe.

With a robust missile defense program, developed under the management of the Missile Defense Agency and under the policy and guidance of the Secretary of Defense, with adequate funding, the United States could have the means to defend the homeland from a range of missile threats in the future. Secretary Rumsfeld in his memorandum of January 2nd has established that the Missile Defense Agency should establish a single program to develop an integrated system. This was a key priority of the program. The company documentation states that among the top priorities is deployment of systems as soon as practicable.

The Patriot Advanced Capability–3, the PAC–3 system, which serves as our first line of defense against short range missiles, is highlighted as an example of this initiative, the first priority listed for missile defense to be used to defend U.S. deployed forces, allies and friends. With the barriers of the ABM Treaty removed, the newly formed Missile Defense Agency is now able to explore the role of defense technologies to defend homeland against all ballistic missile threats, both long range and short range.

Now let me briefly describe a short range threat to the United States. While the tragic events of September 11th have understandably focused our national security attention on the safety of our air space, other vulnerabilities to our homeland are at least as compelling. Though a much beleaguered and highly scrutinized air traffic control system exists today, there is no comparable, comprehensive open ocean ship tracking and identification system that exists for the United States. We do have a 200 mile economic zone, which gives us control over fisheries and other resources in that zone. But our actual jurisdiction over territorial waters for vessels extends only to 12 miles from our shoreline. Even if we did have complete control of the 200 mile economic zone offshore, there exists thousands of missiles in the world today of the Scud class, Scud A, B, C, D, at least, and several of those, particularly later models, could be launched from a ship outside the economic zone, launched from a ship and still hit our coastline.

As the report of the Commission to Assess the Ballistic Missile Threat to the United States noted 4 years ago, sea launch of shorter range ballistic missiles poses a direct territorial threat to the United States sooner than if a country waited to develop an ICBM for launch from its own territory. Sea launching could also permit it to target a larger area of the United States than would a missile fired from the home territory, at least initially. The national intelligence estimate, published by the National Intelligence Council in December 2001, titled Foreign Missile Developments and Ballistic Missile Threat through 2015, bolsters this argument. The NIE states in part, using such a sea platform would not pose major technical problems. The simplest method for launching a ship borne ballistic missile would be to secure a transporter erector launcher on board a ship and launch the missile from the TEL.

The NIE disclosed that several countries are technically capable of developing a missile launch mechanism to use from forward based ships. A number of countries are known to be developing the capability to launch from the sea. The Commission report mentions India specifically in this regard. In addition, press reports of the same timeframe has the work of the Commission tell of Iran test firing a short range surface to surface ballistic missile from the Caspian Sea in the spring of 1998.

However, we don't have to go that far to find evidence of how easy it is to get and ship a Scud missile. A gentleman named Jacques Littlefield proves that not only can nations obtain them, but essentially anyone can. The Los Angeles Times reported in September 25, 1998, that Mr. Littlefield had imported a Scud missile with its launcher into California on September 2nd. The fully operational Scud-B SS–1C, complete with a guidance system and engine, was manufactured in the former Czechoslovakia in 1985. The only critical parts missing were the fuel and a warhead. The Bureau of Alcohol, Tobacco and Firearms, U.S. Customs and the U.S. Navy launched an investigation as to how the missile could have entered U.S. ports without more flags being raised. Littlefield said he wanted the Scud for his weapons collection. Official reports indicate that the Government believed him, as he had already imported another Scud from the Czech Republic 3 months earlier. And by the way, it is not apparent that these are on shipboard when they are shipped. They can easily be covered or concealed as cargo containers or other forms of cargo.

The possibility of a group or even an individual with ambitions of a maleficent nature covertly sailing a ship or a barge with a missile launch platform into targeting range of key population centers in the United States is certainly practical. Lulling ourselves into thinking that such a surprise attack by one or a few sea launch ballistic missiles against U.S. coastal cities, even if only armed with high explosive warheads, is so unlikely as to be negligible is reminiscent of the U.S. attitudes prior to Pearl Harbor in December 1941, and the Japanese attitudes prior to Jimmy Doolittle's raid on Tokyo in April 1942.

About half the U.S. population lives in cities adjacent to coastal territories. And therefore, a very substantial part of our population could be vulnerable to such an attack. I believe that the Missile Defense Agency today has programs in various states of development that can deal with such a threat, but dealing with them may well involve placing PAC–3s, THAADs or other missile defense systems on our coastal territories, as well as having them available to deploy overseas, and could also involve our Aegis fleet defense ships, which could be in our harbors and our offshore territories to provide further defense, as eventually could the Airborne Laser, if that program is brought to successful conclusion.

I will be glad to discuss these capabilities further with you in the testimony. But I believe my time is up. Thank you.

[The prepared statement of Mr. Graham follows:]

Testimony of
William R. Graham, PhD
Former Director, Office of Science and Technology Policy,
Executive Office of the President

Before the Subcommittee on National Security,
Veterans Affairs, and International Relations
Committee on Government Reform
House of Representatives

16 July 2002

Mr. Chairman and distinguished members of the subcommittee, thank you for inviting me to testify on my views concerning the opportunities that are now available in the missile defense area to enhance our national security. I am testifying today on my own behalf, and not as a representative of any specific organization.

The prospective threat to our security from such long-range ballistic missiles as North Korea's Taepo-dong 2 (TD2) is generally accepted and understood. The U.S. is developing missile defenses to thwart such a threat. However, this is akin to only locking the front door of our house and assuming that we will be safe. The U.S. is also vulnerable to attack from short-range missiles. Because this vulnerability has not received the same attention as that of long-range ballistic missiles, I would like to spend some time today addressing this threat. We need to make sure the back door is locked too before we can adequately ensure the safety of the homeland.

Introduction

The termination of the Anti-Ballistic Missile (ABM) Treaty this past June has finally opened up the opportunity to develop effective missile defense capabilities for those who have waited patiently for the time to come when a full range of engineering options in missile defense could be explored. Gone are the constraints on the U.S. that were crafted for the security environment of the 1970's. This freedom permits the U.S. to prepare for threats to our security from those traditional adversaries that may threaten from far away – and now, to prepare for threats that are more ambiguous, and possibly within short-range of U.S. soil. With a robust missile defense program, developed under the management of the Missile Defense Agency (MDA), the U.S. will have the means to defend the homeland from a range of missile threats.

The MDA Mission

Secretary of Defense Rumsfeld, in the Memorandum of January 2, 2002 that established the MDA, indicated that "establishing a single program to develop an integrated system," was a key priority of our Missile Defense Program. Accompanying documentation to this Memorandum states that amongst the top four priorities for the Department of

Defense is the deployment of systems as soon as practicable. The Patriot Advanced Capability-3 (PAC-3) system, which will serve as "the first line of defense against short-range missiles," is highlighted as an example of this initiative. The first priority listed was for missile defense to be used "to defend the U.S., deployed forces, allies, and friends."

The barriers of the ABM Treaty past removed, the newly formed MDA is now able to explore the role of missile defense technologies to defend the homeland against all ballistic missile threats – both long range and short-range.

Offshore Threats

While the events of September 11th have understandably focused our national security attention on the safety of our air space, other vulnerabilities to our homeland are at least as compelling. Though a much beleaguered and highly scrutinized air traffic control system exists today, there is no comparable, comprehensive ship tracking and identification system that exists for the U.S.

A 200-mile (320 kilometer) economic zone extends out from the U.S. coastline. This zone establishes the U.S. right over fisheries and other resources such as oil and gas. Jurisdiction over U.S. territorial waters whereby the vessels that transverse within this area are subject to our control, extends only twelve nautical miles from the shoreline. Even if the U.S. did have a dependable security net draped across the economic zone, this would be ineffective against many ballistic missiles fired from a sea-launch platform. For instance, with a range of 300 to at least 600 kilometers, depending upon the model, a SCUD could be launched from a ship outside of the economic zone and still hit inside the coastline.

As the report of the *Presidential Commission to Assess the Ballistic Missile Threat to the United States* noted four years ago,

> "sea launch of shorter range ballistic missiles…poses a direct territorial threat to the U.S. sooner than if a country waited to develop an ICBM for launch from its own territory. Sea launching could also permit it to target a larger area of the U.S. than would a missile fired from its home territory…"

The National Intelligence Estimate (NIE) published by the National Intelligence Council in December 2001 titled, *Foreign Missile Developments and the Ballistic Missile Threat Through 2015*, bolsters this argument. The NIE states, in part that:

> "…using such a sea platform would not pose major technical problems…The simplest method for launching a shipborne ballistic missile would be to secure a Transporter Erector Launcher (TEL) onboard the ship and launch the missile from the TEL."

The NIE disclosed that several countries are technically capable of developing a missile launch mechanism to use from forward-based ships; and a number of countries are known to be developing the capability to launch from the sea. The Commission report mentions India specifically in this regard. In addition, press reports in the same time frame as the work of the Commission tell of Iran test firing a short-range surface-to-surface ballistic missile from the Caspian Sea in the spring of 1998.

However, Jacques Littlefield proves that it is not only nations that would be interested, and successful, in acquiring ballistic missiles. The Los Angeles Times reported on September 25, 1998 that Mr. Littlefield had imported a Scud missile with its launcher into California on September 2. The fully operational Scud-B SS-1C, complete with a guidance system and engine, was manufactured in the former Czechoslovakia in 1985. The only crucial parts missing were fuel and a warhead. The Bureau of Alcohol, Tobacco, and Firearms; U.S. Customs; and the U.S. Navy launched an investigation into how the missile could have entered U.S. ports without more flags being raised. Littlefield said he wanted the Scud for his weapons collection. Official reports indicate that the U.S. government believed him as he had already imported another Scud from the Czech Republic three months earlier.

The possibility of a group, or even an individual, with ambitions of a maleficent nature covertly sailing a barge-cum-missile-launch platform into targeting range of key population centers of the U.S. is very real. Lulling ourselves into thinking that a surprise attack by one or a few sea-launched ballistic missiles against U.S. coastal cities, even if armed with high explosive warheads, is so unlikely as to be negligible is reminiscent of U.S. preparations prior to Pearl Harbor in December 1941, and Japanese preparations prior to Jimmy Doolittle's raid on Tokyo in April 1942.

Almost 10 million people live in Los Angeles County. A seemingly benign vessel, perhaps not even in U.S. territorial waters, but instead in waters under Mexican jurisdiction, could slip into range for launching a missile at LA, with little difficulty. A missile launched in this scenario would not need to be very accurate in order to do a great deal of damage, particularly when delivering a nuclear or biological warhead.

According to the 2001 U.S. Census Bureau Statistics, approximately 287.5 million people live in the United States. Some 47.4% live along the east and west coast – and Alaska and Hawaii. This means that nearly half of the U.S. population is potentially vulnerable to an offshore threat.

Missile Defense to Defend the Homeland

Developing the means to thwart this threat is not going to be easy. No single system in development, or close to deployment such as the PAC-3, can provide a comprehensive solution. But rather, with a network of systems, it will be possible to provide the coverage necessary to protect our coastlines from the type of offshore threat just discussed. Operationalizing the second priority called out in the missile defense program direction document, to: employ a Ballistic Missile Defense System (BMDS) that layers defenses to intercept missiles in all phases of their flight (i.e., boost, midcourse, and terminal) against all ranges of threats, is required.

The idea is to increase the footprint coverage of our missile defense systems by networking them together. System capabilities should complement one another and provide greater capability in such areas as surveillance, warning, and attribution when linked together than if they were operating independently. For instance, working together, PAC-3, THAAD, Aegis Radar, and future interceptor systems could provide a formidable defense for vulnerable coastal areas of the U.S.

Nearing deployment-ready status, PAC-3 could provide the same type of protection against short-range ballistic and cruise missiles for the American public as is envisioned for allies and U.S. forces deployed abroad. THAAD, designed to protect against longer-range weapons than PAC-3, would enable an extension of defended footprint against incoming missiles. In Congressional testimony this past April, Lt. Gen. Kadish stated that, "THAAD would protect forward-deployed U.S. and allied armed forces, broadly dispersed assets, and population centers against missile attacks." This capability would clearly also be useful for homeland defense. Networked with Aegis, the invaluable radar system depended upon by the Navy for a comprehensive air and sea defense of a given deployment area, these systems could provide a reliable missile defense solution for the American homeland.

The Homeland Defense Precedent

Acceptance of the visible presence of missile defense systems to enhance the security of the United States may strike some as potentially difficult for Americans to accept. Americans pride themselves on living in a free and open society. However, recent circumstances have demanded a review of security practices in a whole host of situations – from checking in at an airline ticket counter to attendance at July 4th celebrations. Such reviews have, without fail, resulted in the implementation of more overt displays of security forces to both deter threats and to more immediately answer them if they were to erupt. A missile defense network such as that outlined here would serve the same purpose.

It is also worthwhile to note that the United States has past experience deploying defensive systems in and around general civilian areas. One such family of systems, known as NIKE, was designed to protect U.S. cities and military instillations from Soviet bombers if they tried to penetrate American air space. First deployed in the early 1950s, the system included interceptor missiles, launchers, radars, and associated command and control. By 1955, there were approximately thirty-eight operational NIKE-Ajax battalions dispersed around the nation. The NIKE-Ajax was eventually deactivated and replaced by a new interceptor missile, known as NIKE-Hercules, which possessed an improved range, speed, and altitude. For many years during the Cold War, the United States deployed nearly 150 NIKE-Hercules batteries spread out across the country protecting key U.S. cities and military instillations. Notable U.S. population centers protected by the NIKE system included Chicago, Philadelphia, New York, and the Los

Angeles area, which alone was ringed by more than a dozen NIKE sites. The NIKE system remained in service until the mid-1970s.

Missile Defense as a Military Mission

A missile defense system designed to protect the American homeland would serve a military mission, just as the NIKE system did before it. Though the objective of such a system is, of course, commensurate with the role of the Office of Homeland Security, the means to achieve the objective uses uniquely military assets. The missile defense mission is inherently part of the military command structure.

As the command structure for U.S. forces has now been delineated, as of 1 October, the protection of the North American Continent will be the responsibility of the Combatant Commander for North America – or NORTHCOM. Secretary Rumsfeld announced another significant change in two combatant commands to take effect 1 October: Strategic Command, or STRATCOM, and Space Command will be replaced by a single new command. This new command, as a functional command with global responsibilities, will support all of the territorial commands, including NORTHCOM in its homeland defense mission. Ultimately, it will be the NORTHCOM Combatant Command that must ensure the security of its area of responsibility.

Relationship to Homeland Security

The options that may be available to defense planners in the missile defense area will expand significantly as scientists and engineers push to realize the full potential of technologies no longer restricted by the ABM Treaty. MDA is a new organization with new opportunities. Achieving a dependable, flexible, and comprehensive missile defense system will be a critical element of our strategy for defending the American homeland.

Mr. PUTNAM. Thank you, Dr. Graham. At this time, we recognize Mr. Eric Miller, senior defense investigator for the Project on Government Oversight and I believe a former Tampa Tribune reporter, correct?

Mr. MILLER. That's correct.

Mr. PUTNAM. Welcome.

Mr. MILLER. Thank you.

Good afternoon, Mr. Chairman and members of the committee. Thank you for this opportunity to comment on the restructuring of the Nation's missile defense program.

As a politically independent watchdog group, the Project on Government Oversight takes no position pro or con on missile defense. We nonetheless have serious concerns that recent missile defense program changes are not in the best interests of our Nation's security or the U.S. taxpayer. Today we are releasing our report, "Big Dreams Still Need Oversight: Missile Defense Testing and Financial Accountability Are Being Circumvented," indicating that the testing oversight of the Missile Defense Agency could signal a step backward to the often misguided acquisition practices of the 1970's and the early 1980's.

Mr. Chairman, at this time I would like to offer our report for the record.

Mr. PUTNAM. Without objection, it shall be inserted into the record at this point.

Mr. MILLER. Thank you.

As we rush to deploy the missile defense program, we would be wise to consider the lack of financial accountability plaguing a growing number of private corporations, public corporations, actually. Though the Pentagon has increasingly been encouraged to conduct business more like the private sector, in this case we must learn from these mistakes, so taxpayers are not victimized, as many shareholders have been.

Foremost among our concerns with the missile defense program is the fact that Defense Secretary Rumsfeld's January directive opens the door for a broader use of special contractual agreements called other transactions. These types of contracts waive many of the financial oversight requirements of typical contracts for goods or services, with the aim of attracting so-called non-traditional defense contractors. Other transactions allow contractors to avoid taxpayer protections and transparency requirements in the Federal Acquisition Regulation and cost accounting standards. These important protections give the Federal Government the information it needs to ensure fair and reasonable contract prices. Another transactions contract on the other hand can even exempt a defense contractor from undergoing Government audits or providing a Federal contracting agency and Government auditors with access to the contractor's pertinent records.

Unfortunately, two other transactions already have been awarded the Missile Defense Agency since their own memo was penned. Both went to traditional, large contractors when Boeing and Lockheed Martin Systems were awarded sole source, other transaction agreements. Obviously, these are not the intended targets of those agreements.

We are also concerned with changes in access by the Office of the Director of Operational Tests and Evaluation. Prior to the establishment in 1984 of the Independent Pentagon Testing Office, far too many overpriced and under tested weapons systems were being placed in the hands of our fighting men and women. We want to remind you today of some of those notorious past weapons failures in the hope that history won't repeat itself.

We're concerned by a new buy now, fix later acquisition chapter could be in the making. No doubt acquisition officials at the Pentagon would love for Congress and the public to forget notorious weapons like the Sergeant York air defense gun, a mobile, armored anti-aircraft system that was approved for production in 1980, before it was battle tested. Nearly 5 years later, after more than $1 billion of public investment, the Sergeant York gun became such an embarrassment that it had to be canceled.

We also hope you don't forget the story of the Bradley fighting vehicle, an armored troop carrier and scout that was approved for full rate production in 1979, even though the Pentagon knew at the time that the vehicle's armor couldn't protect its occupants from hostile fire. Upgrades and design fixes to the Bradley have since been very costly.

We'd also like to jog your memory about the fast track procurement of the B1B bomber, a very costly aircraft rushed into production during the late 1980's, despite catastrophic engine blade failures, munitions limitations and electronic warfare deficiencies, and the C–5 cargo aircraft, a financial boondoggle once dubbed the notorious granddaddy of Pentagon overruns.

In all of these examples, production decisions were made before DOT&E was created, foreshadowing the potential trouble of returning to a system before independent testing. The bottom line of all this is that thorough testing should not be sacrificed in the interest of expediency, nor should financial contracting transparency be abandoned merely to decrease paperwork. The lessons of history tell us that when this happens, the Nation's fighting men and women, as well as the taxpayers, become the losers.

POGO is a solid supporter of rigorous independent operational testing. We are well aware of the Director of Operational Test and Evaluation's numerous accomplishments over the past 17 years. We want the Director to continue to provide much needed objective analysis. But we are concerned that a new area of secrecy at the Missile Defense Agency will cut him out of the loop on some important aspects of early testing.

Right now, the Director, Mr. Christie, says he has an amicable relationship with the missile agency, and that he's confident it will continue. But what happens if that relationship sours? What happens if he gives the program a bad report card, or presses too hard for data that the agency doesn't want to relinquish? History has shown that such relationships can quickly go south when the facts don't fit the Pentagon story. The Office of the Director of Operational Tests and Evaluation should not be required to negotiate the nature of information it is provided by the missile defense program managers. The Office should have unfettered access and be an active participant in early testing of missile defense systems.

Than you for inviting me to testify before the subcommittee. I am happy to answer any questions.

[The prepared statement of Mr. Miller follows:]

Testimony of
Eric Miller, Senior Defense Investigator
Project On Government Oversight
on
Missile Defense Oversight
Before the
House Government Reform Subcommittee on
National Security, Veterans' Affairs, and International Relations
July 16, 2002

Mr. Chairman and Members of the Committee, thank you for this opportunity to comment on the restructuring of the nation's missile defense program.

The Project On Government Oversight (POGO) investigates, exposes, and seeks to remedy systemic abuses of power, mismanagement, and subservience by the federal government to powerful special interests. Founded in 1981, POGO is a politically-independent, nonprofit watchdog that strives to promote a government that is accountable to the citizenry.

While we take no position, pro or con, on missile defense, we nonetheless have serious concerns that recent missile defense program changes at the direction of Defense Secretary Donald Rumsfeld are not in the best interest of our nation's security or the U.S. taxpayer. Today, we are releasing our report, Big Dreams Still Need Oversight: Missile Defense Testing and Financial Accountability Are Being Circumvented. It illustrates why and how reductions in financial and testing oversight of the Missile Defense Agency could signal a step backward to the often misguided acquisition practices of the 1970s and early 1980s.

As we rush to deploy the missile defense program we would be wise to consider the lack of financial accountability plaguing a growing number of corporations. Though the Pentagon has increasingly been encouraged to conduct business more like the private sector, in this case, we must learn from these mistakes so taxpayers are not victimized as many shareholders have been.

We are concerned that Secretary Rumsfeld's directive opens the door for a broader use of special contractual agreements called "other transactions." These types of contracts waive many of the financial oversight requirements of typical contracts for goods or services with the aim of attracting so-called "nontraditional" defense contractors. "Other transactions" allow contractors to avoid taxpayer protections and transparency requirements in the Federal Acquisition Regulation and Cost Accounting Standards. These important protections give the federal government the information it needs to ensure fair and reasonable contract prices. An "other transactions" contract, on the other hand, can even exempt a defense contractor from undergoing government audits or providing the federal contracting agency and government auditors with access to the contractor's pertinent records.

Unfortunately, two "other transactions" have already been awarded by the Missile Defense Agency since the Rumsfeld memo was penned. Both went to traditional, large contractors when

Boeing and Lockheed Martin Mission Systems were awarded sole-source "other transaction" agreements. Obviously, these are not the intended target of these agreements.

The Director of the Missile Defense Agency should be required to follow the intent of the law permitting "other transactions." The Director should only be permitted to negotiate such agreements with nontraditional defense contractors that would otherwise not offer their expertise.

Prior to the establishment in 1984 of an independent Pentagon testing office – the Director of Operational Test and Evaluation – far too many overpriced and under-tested weapons systems were being placed in the hands of our fighting men and women. We are here today to remind you of some of those notorious past weapons failures in the hope that history won't repeat itself. We're concerned that a new "buy now, fix later" acquisition chapter could be in the making. Not only does the Rumsfeld directive potentially weaken the mission of DOT&E, it also will allow Missile Defense Agency officials to essentially waive financial contracting requirements that could allow millions of dollars in missile defense spending to disappear into the Pentagon's financial management black hole.

Frankly, one of our purposes for being here today is to bring back a few unpleasant memories. No doubt, acquisition officials at the Pentagon would love for Congress and the public to forget notorious weapons like the Sgt. York Air Defense Gun, a mobile, armored anti-aircraft system that was approved for production in 1980 before it was battle-tested. Nearly five years later, after more than a billion dollars of public investment, the Sgt. York gun became such an embarrassment that it had to be cancelled.

We also hope you don't forget the story of the Bradley Fighting Vehicle, an armored troop carrier and scout that was approved for full-rate production in 1979 even though the Pentagon knew at the time that the vehicle's armor couldn't protect its occupants from hostile fire. Upgrades and design fixes to the Bradley have since been very costly.

We'd also like to jog your memory about the fast-track procurement of the B-1B Bomber, a costly aircraft rushed into production during the late 1980s despite catastrophic engine blade failures, munitions limitations, and electronic warfare deficiencies; and the C-5 cargo aircraft, a financial boondoggle once dubbed the "notorious granddaddy of Pentagon overruns."

In all of these examples, production decisions were made before DOT&E was created, foreshadowing the potential trouble of returning to a system before independent testing. The bottom line of all this is that testing should not be sacrificed in the interest of expediency, nor should financial contracting transparency be abandoned merely to decrease paperwork. The lessons of history tell us that when this happens the nation's fighting men and women, as well as the taxpayers, become the losers.

POGO is a solid supporter of rigorous independent operational testing. We are well aware of DOT&E's numerous accomplishments over the past 17 years. We want the Director to continue to provide much needed objective analysis. We are concerned that a new era of secrecy at the Missile Defense Agency will cut him out of the loop on some important aspects of early testing.

Right now Mr. Christie says he has a amicable relationship with the agency, and that he's confident it will continue. But what happens if that relationship sours? What happens if he gives the program a bad report card, or presses too hard for data that the agency doesn't want to relinquish? History has shown us that such relationships can quickly go south when the facts don't fit the Pentagon's story.

The Director, Operational Test and Evaluation should not be required to negotiate the nature of information it is provided by missile defense program managers. The office should have unfettered access, and be an active participant in early testing of missile defense systems.

In addition, Congress mandated that the Pentagon successfully complete DOT&E's operational testing before deploying a national missile defense system. I cannot stress the importance of holding them to this requirement.

Thank you for inviting me to testify before the Subcommittee. I am happy to answer any questions.

# # #

Mr. PUTNAM. Thank you very much.

Looks like it's just you and me, babe, we're going to give you 10 minutes and start us off right.

Mr. TIERNEY. I don't have to take that long. Mr. Miller, I want to thank you for the work that you've done and your organization has done on that. I was also disturbed to listen to Mr. Christie play what I think is a lapdog, basically, to the Pentagon here. I was very disturbed to hear him say that he hasn't got all the information he needs yet, but he hopes he's going to get it and thinks maybe he will, that he understands he has nothing to measure against performance, but that he's going to take a look at what's there.

Can you give us a little bit broader definition of what it is that you think the importance of the Office of Testing and Evaluation for early involvement in these programs, and what that means to the overall savings of money to people, but also savings of lives?

Mr. MILLER. I realize that development of testing is the purview of the Service. But it's very important that the Operational Test and Evaluation, the independent tester be involved early on to sort of get a system ready for testing to find out if it can be tested. There are frequent problems, anything from a system not really being ready to be tested to not having enough, say, if you were going to test an aircraft, there may not be enough aircraft to adequately test it.

So the Director of Operational Test and Evaluation needs to be involved, needs to monitor what's happening with the program, so that he knows what he can look at when he is ready to test it to give it sort of its final exam.

Mr. TIERNEY. When Mr. Christie tells us he's going to get some of his information from this new advisory committee that General Kadish has set up, essentially I would suggest that isn't enough, that his people have to be right in there, getting the raw data and watching the tests first hand in order to be fully informed. Would that be accurate?

Mr. MILLER. Yes. What concerns us is that there shouldn't even be a need for this paperwork to go between the offices, that Mr. Christie, we believe by law, has a right to all the testing data. And sometimes, even if he's getting what he's asking for, there might be some things that he doesn't know about, and it might come back to actually jeopardize the outcome of the weapons system if he doesn't know what to ask for.

Mr. TIERNEY. You mentioned a minute ago that one of the problems that we have on this is that if the relationship falls out or the military just decides that they're not going to share something with him, one of the reasons Mr. Coyle looked into the report was that the President specifically asked him to do an evaluation at that point in time. He did an evaluation that was extensive, and came up with 50 recommendations of things that needed to be changed.

I suggest, Mr. Chairman, that may have been the stimulus for this entirely new revamping of the system to where now there won't be any more evaluation at that stage, where we've got a whole new nomenclature for what's going on, and we're not getting those criticisms addressed and everything is going to be classified so there won't be any more criticism. That's the reason concern, at least from this one Member, if it goes forward on that. I think that Mr. Christie makes a good point in his testimony and in his report, if you look at it, as to what can happen if we don't maintain the rights of Mr. Christie to step in there at an early stage and see what's going on, and the obligation of Congress to look at that from time to time against some benchmarks that can help us evaluate it.

It's one thing, Mr. Miller, maybe you can tell us, are you aware of a program in the past where you've had no benchmarks to evaluate anything against at all, it was just basically take a look at it from time to time, and if we like where it's going, as a Pentagon, we'll keep on going?

Mr. MILLER. I can't recall any major weapons system program. I know there are a lot of experimental types of programs where they don't even have independent operational——

Mr. TIERNEY. But that's below a certain dollar amount, right?

Mr. MILLER. I haven't run into any major programs myself, in my personal experience.

Mr. PUTNAM. Would the gentleman yield?

Mr. TIERNEY. Sure, I'll yield.

Mr. PUTNAM. For my benefit, maybe to stimulate some dialog for where you're headed as well, I would encourage other witnesses to jump in there as well.

Mr. TIERNEY. Well, if they're qualified. I don't know. Mr. Levin may be, I don't know if the others are qualified for this area. But if they want to.

Mr. PUTNAM. Dr. Graham.

Mr. GRAHAM. Thank you. I am a physicist and engineer who's been involved in one way or another with military hardware since I was a lieutenant in the Air Force. So at least I have some credentials going back on this subject.

It depends on the state of the technology that you are pursuing as to what your aspirations are. As General Kadish tried to explain, aircraft are now about 100 years old. The Wright brothers did not have a carefully crafted set of criteria to which they subjected the first flyer, even though the military did try to buy an airplane from them soon thereafter.

Obviously today we know a lot about airplanes and a lot about how to describe the characteristics we want. An example, perhaps closer to the ballistic missile defense challenge today would be the Manhattan project, where the goal was to build an atomic bomb, whatever that was. And the approach taken to it was multi-faceted. Enriched uranium bombs, plutonium bombs, gun bombs, implosion bombs, and so on, we tried a lot of things. And it turned out, unfortunately in some ways, it was easier than the physicists and engineers working on it thought.

Over time, as we've become more sophisticated and knowledgeable about it, we've learned how to give more specific direction to the developers and to test them to some extent, although I'm afraid to tell you today we're not able to test our nuclear weapons stockpile any further. But there are a couple of examples of places where we don't have specific tests in mind when we began the development process.

Mr. TIERNEY. One of the issues they have with that is that this project is about 8 to 10 times more expensive than the Manhattan Project is going to be, using today's dollars. And the other is that there are parts of this technology that certainly could be tested. If you throw it all together, as they have, it's a very clever way of throwing it all together and saying, we've never been to the end of this road before, but there are steps along the way and technologies along the way that ought to have benchmarks for us, not only as they are working individually, but as they are working in combination. That's where part of the problem comes on this thing.

Mr. GRAHAM. As an engineer, I'd say that testing is only one manifestation of the development program. It's a very important manifestation, but in fact you start out with the designs, you do calculations and you do simulations, then you do hardware in the loop tests, you evaluate it through all of that and finally, you run various types of tests. Most of which you do for engineering purposes to see if your models are accurately predicting the performance that you're achieving in the real world.

Mr. TIERNEY. Without having any models to test that, right?

Mr. GRAHAM. The operational evaluation test is the final step in that, and in my mind, one of the less important ones.

Mr. TIERNEY. But I think you hit it on the head when you said you had models that you were evaluating to see whether or not you were at that point in time. And that, I guess, comes back to our problem here, we're not going to be given those points of measurement and things to evaluate it against.

Mr. GRAHAM. There are a considerable quantity of models and simulations and hardware in the loop tests and facilities that are engaged daily in the ballistic missile defense program. I'm sure that, while I can't speak for General Kadish, I believe he would be more than pleased to have you and your staff visit those facilities and inspect them.

Mr. TIERNEY. You're right, you can't speak for General Kadish, because he hasn't made that offer and he hasn't made those available. I think that's the reason we're having this hearing today.

Mr. LEVIN. Mr. Tierney, can I add a perspective here?

Mr. TIERNEY. Sure, please do, Mr. Levin.

Mr. LEVIN. The General Accounting Office and TO&E are in lots of ways very similar as being outside critics of Missile Defense Agency and its activities. I think there is a natural healthy tension there between the parties. And you don't want everybody just nodding their heads, yes, everything's fine, everything's fine, and have the Missile Defense Agency come out and say, trust us, we know what we're doing, we're making progress, don't you see the tests, we intercepted everything just fine, you know, just keep giving us $7 billion.

I mean, the perspective is, it doesn't hurt to have somebody looking from the outside and making suggestions. We did that in our Airborne Laser work for Chairman Shays and this committee. I think if the Missile Defense Agency adopts those recommendations, they'll make progress. Just like they're making progress in their testing program, because they're accepting the recommendations of Phil Coyle and TO&E that were made back in August.

So there's a healthy tension, I think that's good.

Mr. TIERNEY. Well, I think that's what we aim to continue. The problem is that under this new plan, I don't think we're going to have the opportunity to have a Phil Coyle or a GAO get the information they need to make those evaluations. I don't know if you're the person from GAO that could testify to it or not, but GAO has had considerable amount of difficulty getting information. Is that your experience, or somebody else from your office?

Mr. LEVIN. Oh, I'm the person. I'm the person to talk to.

Mr. TIERNEY. Well, and that's my point. We want that tension to be there, then people have to play straight up and fair.

Mr. LEVIN. I take General Kadish and Mr. Christie at their words that there is now unfettered access. I'm not convinced that there was unfettered access before questions were being raised by yourself and others.

In terms of GAO's access to date, we have had concerns. I can't say we've totally resolved those concerns. We are working with MDA, in fact, we have an ongoing task force, task group that meets about every other week to figure out what GAO can do to get better, put in better requests so MDA understands what we're asking for. And also what MDA can do to streamline its processes. We would like to be able to go to a meeting and say, please hand over the document right now. They feel like they need to take a look at what they're giving us internally. We've had problems getting things in a rapid fashion in the past, and so we've raised this level of concern all the way up to General Kadish. And they're trying to satisfy us, we're just not there yet.

Mr. TIERNEY. And just to let you know, in Mr. Christie's testimony, he indicates that he's not yet involved in the other programs, other than the PAC–3 program, he is not involved the way he wants, and he says he's not there yet, but he wants to be. That's the issue, the same thing with your office, according to the information that we have. I yield back.

Mr. PUTNAM. Thank you, Mr. Tierney.

I'd like to direct my question to Ambassador Smith. You have a background in the international treaties and certainly working with the ABM Treaty. We're involved on an international basis today with Meads and Ramos and Arrow and SM–3 and some other things. Are there restrictions under INF and START that are any way a hindrance to the development of National Missile Defense?

Ambassador SMITH. Mr. Putnam, of course the chief impediment treaty was the ABM Treaty. But the fact is, there are some problems that we have that remain with the INF Treaty and the Start Treaty. The problems that arise are generally with creating good emulating target missiles. We have a situation where most of the missiles we worry about in the world are somewhere between the range of 550, 500 kilometers. China has them, Pakistan has them, Iran has them, we can go on with a long list of countries that can have them, want them.

The two countries that can't have them are the United States and Russia. Unfortunately, up until now, that has also meant creating a modern target missile. So when an engineer goes to do the kind of realistic testing that is being demanded, the kind of realistic testing that I think MDA would like to do, he runs into that kind of a barrier. You run into similar problems with the START

Treaty if you want to do this as a surface sea launch or an air launch. That is, if it has a range of greater than 600 kilometers, and you want to launch it from a surface ship, or you want to launch it out of an airplane, you can't do that under START.

Now, the good news here is unlike the ABM Treaty, which frankly was something that simply was something of the past and had to be superseded, there was no way to fix it, I think if the will were there politically that these are the kinds of things we could reach either some amendments or understanding with the Russians. We don't need to attack these treaties wholesale. But the fact is, if you want a good, modern emulating the threat target missile, we need to do something about the INF Treaty and the Start Treaty.

Mr. PUTNAM. Dr. Graham, you mentioned that the greatest threat, you believe, that is out there is a sea based, short range missile, perhaps even launched within our economic territory. Based on what Ambassador Smith has said, how do we prepare appropriate countermeasures, how do we adjust our homeland security plans, which is all the rage here in Congress right now? How do we prepare for that type of an attack?

Mr. GRAHAM. Well, right now, much of the country is in the same frame of mind that the population of the Hawaiian Islands were on December 6, 1941, believing that such an attack couldn't possibly take place. Unfortunately, those of us who are engineers and scientists who have worked in the defense area know that several countries have launched ballistic missiles from the decks of ships, the United States, Soviet Union, Iran, India and other European countries. It is not difficult to do.

While North Korea may have a few Taepo-dong 2 missiles, there are literally thousands of Scud missiles being built by tens of countries in the world today. And by my example of the collector who bought two of them, I tried to show how easily they can be obtained.

So in the approach that says, don't give your enemies any free shots at you, it seems to me that we should take an interest in developing or in deploying defensive capability against the near offshore threats, just as we are developing capability against the long distance missile threats. Putting PAC–3 near coastal cities would give those cities not only short range ballistic missile defense but also cruise missile defense and hostile aircraft defense. That's quite feasible, the PAC–3 is coming into the inventory very shortly. In the past, the United States has had many city located ground based anti-aircraft systems deployed in their proximity. I think Los Angeles alone had about a dozen Nike sites back in the 1950's and 1960's, and most large U.S. cities did, too. So we may have to move in that direction and augment that with other systems like THAAD, Aegis, and Aegis when they come on line.

I think it's eminently doable. Our problem isn't our technical ability to intercept such threats. Our problem is our perceptual ability to conceive that such threats could threaten us.

Mr. PUTNAM. What type of platform requirement is there? Could it be launched from something that's camouflaged as a commercial fishing vessel? What are we involved in there? In other words, how difficult is it, if it's easy for a collector to buy a Scud, how easy is it for him to actually acquire the warhead, the fuel and the ability to launch that, if you were a nonstate actor?

Mr. GRAHAM. It's not difficult at all. In fact, the world market is awash with that equipment, and you can buy it as a private individual, you can buy it as a transnational group, you can buy it as a small country. And in addition to that, it can be deployed on relatively small ships, small tramp steamers, medium size fishing boats. You can disguise it as a couple of cargo containers until such time as you want to erect the missile and launch it. That too is not difficult.

So it's available, the fuels are available, the techniques for launching them are widely known, and there are many trained people in the Eastern Bloc and elsewhere who have acquired that skill. It's just nothing that has large barriers to keep it from being done, should some group want to do that.

Mr. PUTNAM. Do you believe that's a greater threat than the longer range threat?

Mr. GRAHAM. I believe that there are more short range missiles that could be launched that way today than there are long range missiles, if you exclude Russia and China. But more fundamentally than that, I believe that if we neglect and completely ignore any aspect of our defense, be it from cruise missiles, aircraft, ballistic missiles, long or short range, if we neglect any aspect of that, we're inviting terrorists, adversaries, to take advantage of that neglect to attack us. And that has been the story of our being attacked in the past. After all, if the United States is thinking about a threat and preparing itself for it, we are very difficult to beat. But when the United States doesn't conceive of the threat and pays no attention to it, that's when we find ourselves vulnerable.

Mr. PUTNAM. Ambassador.

Ambassador SMITH. Mr. Putnam, if I may just reinforce what Dr. Graham has said. This worldwide market out there is extremely active. The editor of Jane's Rockets and Missiles has actually traced the collusion between Pakistan and Iran and North Korean in building these things. I think it's no accident that a lot of the countries we're talking about also happen to find themselves on the State Department's list of state sponsors of terrorism.

We tend to think of a state with a missile and we sort of mirror image the kinds of protections we have here, or even to a slightly lesser extent but nevertheless real the kinds of protections the Russians have. The fact is, what we're finding now as we look at the situation in the Pakistan-Afghanistan area, there are things we didn't know about Pakistan, or at least most Americans didn't know. There are entire areas called the border areas and the tribal areas that are simply not administered by the Federal Government of Pakistan. We have the Inter-Services Intelligence Agency that was in collusion with the Taliban in Afghanistan. It is still extremely powerful and one of the biggest problems that General Musharef has trying to get his hands around this.

The notion that somehow a missile or a warhead or both are cobbled together from here or there could somehow slip out of the arsenal of a state and into the hands of a fairly well organized terrorist group is not all that fantastic. Anyone who wants to scoff at this should take a look at what we were thinking before September

11th. That was the stuff of Tom Clancy. This is not all that far-fetched. The notion that somebody could have a handful of these on a ship is really not that far-fetched.

I'd also like to comment about the notion of a long range missile, because it seems to me that the nearer term threat and the greater numbers are exactly as Dr. Graham suggests. There are more Scuds out there that you could put on a ship. If you wanted to use a missile against the United States, you put a Scud on a ship. In the longer term, countries are working on things like ICBMs. I think there you get into a different sort of threat, and that's the threat of geopolitical blackmail, keep the United States out of my area. That's a different threat, but nevertheless also one that is real that we have to protect against.

Mr. PUTNAM. Mr. Miller, you said that thorough testing should not be sacrificed for expediency. When would it be expedient to sacrifice thorough testing, if ever? In time of war? And that's open for anyone on the panel.

Mr. MILLER. I'm not exactly sure at which point it would become expedient. But I think as a broader question, you might have to ask what program should be exempt from traditional testing and oversight. I'm not sure it would be based strictly on immature technology, because I think although the technology and missile defense may be immature, and I'm not an engineer but it's been in development since probably 1983.

And it's been around a while, and I know there are numerous weapons systems that have had immature technologies in various aspects, for example, the Crusader, which was recently canceled, or may be canceled, or recommended for cancellation. They had a benchmark of having a liquid propellant, they tested it, the liquid propellant didn't work. But I'm not sure that just because you have an immature technology that you don't have benchmarks.

Mr. PUTNAM. Does anyone else want a stab at that?

Mr. LEVIN. What we're trying to see MDA do, to a much greater extent than they're doing now, is using technology readiness levels to measure the maturity of the technology, so you understand exactly how representative your hardware is and how representative and how realistic the environment is that you're testing your equipment in. So what we would like to see is a greater emphasis on using the TRLs to manage the program. These are the kinds of benchmarks that I think provide the transparency and accountability and understanding of progress that aren't there today.

You might reach a point where you feel like you have to put emergency capability into place because of the threat. It's good to know exactly what the limits are of your capability. And you might make a decision, well, despite the limits, I'm going to put it into operation, because of the threat. But it's good to know what those risks are.

But under normal, non-emergency situations, I mean, you want to have a disciplined process for developing the technology. That's what we're looking for.

Mr. PUTNAM. I'm glad that you mentioned the TRLs. We borrowed that from NASA. To me there's a lot of parallels between NMD and the space program. When we launched the space program, were there milestones, were there benchmarks other than we will put a man on the moon by the end of this decade?

Mr. LEVIN. I'm not that familiar with how NASA used TRLs. Although like you said, NASA did develop the TRL metric.

What we have learned, certainly, is that you can't try to set schedules, like getting a man on the moon or intercepting a ballistic missile at 2,400 kilometers by a certain date, with an operational system, unless you've done the hard work of development and testing and built your TRLs from what, in the case of Airborne Laser and many other missile defense systems is still pretty immature. You have to be able to build over time, build and test and see what your capability is. And then what GAO recommends is you reach what we call knowledge point one. It's kind of hard to see on the slide, I hope you have it in front of you. That's where you'd come to a match between what your technology can do and how it's proven out and how realistic it is and what the capability you need and the requirements are. You're supposed to freeze the requirements at that point. That's when you can set the cost and schedule.

Until you reach that match between the requirements and the available resources, which is the technology, the time and the money, your estimates are going to be very unrealistic. That's what we saw in the Airborne Laser, where the program was 50 percent over budget and 4 years behind schedule.

Mr. PUTNAM. Mr. Tierney, did you have any other questions?

Mr. TIERNEY. No, Mr. Chairman. I'm aware that there's another committee that wants this room and is entitled to it. I want to thank all the witnesses for their time, for listening to the first panel as long as it took and for your testimony today.

Thank you.

Mr. PUTNAM. We appreciate your input very much. Is there anything very briefly that has gone unsaid that you have been sitting there all morning waiting to be asked and haven't had the opportunity to give us anything? Anyone? Dr. Graham, very briefly.

Mr. GRAHAM. Thank you, Mr. Chairman, very briefly. I would say a great deal has been made about the reality of testing. I would say that testing as an engineering discipline is a much richer subject than just a reality or not reality characterization. You will never have a test so real that it is a missile under the control of a hostile country launched from that hostile country on their preferred trajectory at us. So in that sense you will never have a fully realistic test.

On the other hand, as an engineering discipline, a lot is done about predicting how interceptors as well as the targets fly before they are ever launched. And the real meat of the subject is in the comparison of how the systems we're developing perform against the models and simulations we do and then against the models and simulations of the threat.

One last comment. In most cases, we will know more about countermeasures enemies are using against us than they do, because if they test them, we have far better sensors to observe their performance than the enemy does, both space based, air, ground and sea based sensors. If they don't test them, they will have very little confidence in their ability to perform.

Mr. PUTNAM. Anyone else? Mr. Miller.

Mr. MILLER. I would just encourage you as guardians of the public interest to continue to monitor the financial and testing aspects of the program. It is one of the most expensive ever weapons systems development in the history of the country. And we would sure hope you would keep an eye on it. Thank you.

Mr. PUTNAM. We will, and I know Mr. Tierney will. And we appreciate all of your efforts and everyone's participation. With that, the hearing stands adjourned.

[Whereupon, at 2:37 p.m., the subcommittee was adjourned, to reconvene at the call of the Chair.]

○

APEX
TP
9780160695162
76901945 – 88